T0228093

Harmful Algae Blooms in Drinking Water

Removal of Cyanobacterial Cells and Toxins

Advances in Water and Wastewater Transport and Treatment

A SERIES

Series Editor
Amy J. Forsgren

Xylem, Sweden

Harmful Algae Blooms in Drinking Water: Removal of Cyanobacterial Cells and Toxins
Harold W. Walker

ADDITIONAL VOLUMES IN PREPARATION

Harmful Algae Blooms in Drinking Water
Removal of Cyanobacterial Cells and Toxins

HAROLD W. WALKER

CRC Press
Taylor & Francis Group
Boca Raton London New York

CRC Press is an imprint of the
Taylor & Francis Group, an **informa** business

CRC Press
Taylor & Francis Group
6000 Broken Sound Parkway NW, Suite 300
Boca Raton, FL 33487-2742

First issued in paperback 2017

ISBN-13: 978-1-4665-8305-4 (hbk)
ISBN-13: 978-1-138-74945-0 (pbk)

Library of Congress Cataloging-in-Publication Data

Walker, Harold W., 1968-
  Harmful algae blooms in drinking water : removal of cyanobacterial cells and toxins / author, Harold W. Walker.
      pages cm
  Includes bibliographical references and index.
  ISBN 978-1-4665-8305-4
  1. Water--Purification--Microbial removal. 2. Algae--Control. 3. Cyanobacterial blooms--Prevention. I. Title.

  TD427.C92W35 2015
  628.1'62--dc23   2014025649

Contents

Preface

The increasing presence of harmful blooms of cyanobacteria is a major challenge facing water managers and drinking water utilities across the world. In recent years, for example, blooms of *Microcystis* sp. have raised concerns as these cyanobacteria secrete a class of toxins called microcystins. Microcystins are known liver toxins and skin irritants and are on the US Environmental Protection Agency (USEPA) Contaminant Candidate List (CCL). The World Health Organization (WHO) established a provisional guideline value of 1 µg/L for microcystin in drinking water. Additional cyanobacterial toxins that threaten drinking water include cylindrospermopsin, antitoxin, saxitoxin, and others.

Traditional drinking water treatment processes, including coagulation, softening, filtration, and disinfection, are generally ineffective at removing all cyanotoxins from drinking water but may effectively remove intact cells under optimal conditions. Of concern, however, is the fact that coagulation, flocculation, and disinfection can break open cell walls and release toxins during treatment if not configured properly. As a result, drinking water treatment plants must consider the operation of existing processes as well as the construction of costly additional unit processes to remove cyanotoxins. Possible advanced processes for removing cyanotoxins include granular activated carbon (GAC), powdered activated carbon (PAC), membrane processes, or advanced oxidation processes.

The primary aim of this book is to synthesize the current research literature and provide a description of the factors affecting the occurrence and treatment of cyanobacterial cells and toxins in drinking water. The chapters that follow provide the necessary background so that treatment plant operators, engineers, and water managers can understand the vulnerability of drinking water treatment plants to cyanotoxins as well as to develop treatment schemes to minimize the impact of these contaminants. This book will also be valuable to researchers studying the occurrence, toxicity, and treatment of cyanobacterial cells and toxins.

About the Author

Harold Walker is director and professor of civil engineering at Stony Brook University. Prior to coming to Stony Brook, Dr. Walker was a professor in the Department of Civil, Environmental, and Geodetic Engineering at The Ohio State University. Dr. Walker also served as the director of the Ohio Water Resources Center, the federally authorized and state-designated water resources research institute for the state of Ohio. Dr. Walker served on the board of directors of the National Institutes for Water Resources as well as on the board of directors for the Water Management Association of Ohio. Dr. Walker is a registered professional engineer in the state of Ohio. He has a BS in environmental engineering from Cal Poly San Luis Obispo and MS and PhD degrees in civil and environmental engineering from the University of California, Irvine.

Dr. Walker's research focuses on understanding surface chemical processes in natural and engineered systems, with an emphasis on clean water. Current areas of research include predicting the fate and transport of manufactured nanomaterials, cyanotoxins, and other "emerging" contaminants in groundwater, lakes, oceans, and water treatment plants; developing novel membrane treatment systems and membrane cleaning approaches; and determining the vulnerability of the public to cyanotoxins in finished drinking water. Dr. Walker has secured external funds to support his research from a number of sources, such as the National Science Foundation (NSF), US Environmental Protection Agency (USEPA), United States Geological Survey (USGS), Department of Energy (DOE), National Oceanic and Atmospheric Administration (through Sea Grant), and a variety of state agencies, industries, and other sources.

1

Introduction

1.1 Harmful Algal Blooms and Harmful Algal Bloom Toxins

Harmful algal blooms (HABs) occurring in freshwater, and the associated toxins they produce, are dangerous to animals and humans. HABs also reduce the recreational value of surface waters, thereby influencing tourism and the quality of life for residents and visitors. A number of reports and studies have been carried out since the beginning of this century to elucidate the extent of the problem in freshwater systems, our current understanding, and future research needs [1–4]. Many organisms may produce freshwater HABs, including cyanobacteria (formerly known as blue-green algae), haptophytes, macroalgae, euglenophytes, dinoflagellates, and others. Cyanobacteria are typically the taxa of greatest concern with respect to HABs in freshwater systems and include organisms such as *Anabaena*, *Microcystis*, *Nodularia*, *Cylindrospermopsis*, and others. Not all cyanobacteria are toxic. In some cases, species known to produce toxic metabolites do not do so. Research suggests, however, that the majority of cyanobacterial blooms observed are toxic in nature.

Cyanobacteria and other HAB organisms produce a variety of toxins, including neurotoxins, hepatotoxins, dermatotoxins, and other bioactive compounds. As shown in Table 1.1, the most typically cited classes of HAB toxins include microcystins, saxitoxins, anatoxins, and cylindrospermopsin. Microcystins, for example, are a class of HAB toxins that consists of dozens of congeners with slightly different structure and toxicity. Related compounds to saxiton (STX) include neosaxitoxin (NSTX), gonyautoxins (GTX), and decarbamoylsaxitoxin (dcSTX). HAB toxins are known to cause a variety of short- and long-term health effects in animals and humans. In natural systems, HABs result in fish kills and may lead to death or acute health effects in animals and livestock. Short-term, acute effects of HAB toxins on humans include rashes, liver inflammation, numbness, dermatitis, gastrointestinal problems, liver failure, and others. Documented longer-term impacts associated with low-level, chronic exposure to HAB toxins are less well known but potentially include tumor formation, cardiac arrhythmia, and liver failure.

TABLE 1.1

Common Classes of HAB Toxins

Toxin	Abbreviation	Congeners or Related Compounds	Toxicity	MW
Microcystin	MC	>80	Liver	~1000
Cylindrospermopsin	CYN		Liver, kidney	415
Anatoxin	ANA		Neurotoxin	165
Saxitoxin	STX	NSTX, GTX, dcSTX	Neurotoxin	~300

FIGURE 1.1 (See color insert.)
Satellite image of the western basin of Lake Erie in summer 2011. The image shows a massive harmful algae bloom reaching from the southern to the northern shore. (Image courtesy of NOAA.)

In 2011, a massive bloom of *Microcystis* occurred in the western basin of Lake Erie, as can be seen in the satellite image in Figure 1.1. The bloom occurred in early October and was believed to cover an area of nearly 3000 km². The bloom resulted in extremely high levels of the toxic compound microcystin. In some areas, the concentration of microcystin was measured in the hundreds of micrograms per liter. The bloom was suspected to be at least partially caused by unusually heavy rain the previous spring, which increased the loading of nutrients to the lake [5]. In August of 2014, a smaller *Microcystis* bloom on Lake Erie occurred near the water intake structure for the City of Toledo water supply and left over 400,000 people without drinking water.

Cylindrospermopsin is another cyanobacterial toxin of increasing concern. The first documented outbreak of *Cylindrospermopsis* occurred in 1979 on Palm Island, Queensland, Australia [6]. The outbreak (or so-called mystery disease) occurred following the application of copper sulfate to control an algal bloom in Solomon Dam and resulted in the hospitalization of over 100 aboriginal children. It is believed that the addition of copper sulfate disrupted *Cylindrospermopsis,* which facilitated the release of the toxin. Cylindrospermopsin is a potent toxin to the liver and kidney.

Anatoxin-a was first discovered in the 1960s after an outbreak of *Anabaena* floc aquae in Saskatchewan Lake, Canada, resulted in the death of several cattle. Anatoxin-a is also referred to as "very fast death factor" because of its potent neurotoxicity. Anatoxin-a leads to convulsions and respiratory paralysis. Saxitoxins are another class of neurotoxic compounds produced by HABs in both marine and freshwater systems. Saxitoxins are well known as the cause of paralytic shellfish poisoning.

1.2 Causes of HABs

The causes and factors controlling the formation and occurrence of HABs are varied and complex. As mentioned, the historical HAB in Lake Erie in 2011 is thought to be largely a result of climatic conditions the previous spring. Thus, hydrologic conditions are thought to play a major role in promoting the occurrence of HABs. The flow into and out of a surface water body, weather conditions, timing of rain events, the depth of the water body, water temperature, and the extent of stratification can all influence the timing and extent of HABs. Another important cause of HABs is the input of nutrients into a reservoir or lake. Nitrogen, phosphorus, and other nutrients limit algal growth in most cases, such that when increases occur in these limiting nutrients, rapid growth of algal species follows.

Generally, phosphorus is the limiting nutrient in freshwater systems; nitrogen may limit the growth of algae in marine or estuarine systems. Nutrient inputs come from many sources, including the atmosphere (e.g., nitrogen oxides); point sources (e.g., wastewater treatment plants, industrial sources); and nonpoint sources (e.g., agricultural, on-site wastewater treatment systems). One of the most important nonpoint sources of nutrients comes from the application of fertilizer, for either large-scale agricultural production or residential lawn care. On-site wastewater treatment systems, such as cesspools and septic tanks, can also be important sources of nitrogen to surface water if transport of nutrients through groundwater to surface water is favorable. Internal sources of nutrients can also be important in the formation and maintenance of HABs, especially when conditions are favorable for the resuspension of nutrients previously inaccessible in bottom sediments.

Ecological factors also play a role in controlling the timing, magnitude, and extent of HABs. A particular surface water, for example, may be more prone to HABs in years following a bloom event. In such cases, HAB taxa may "winter" in bottom sediment and be reintroduced back into the water column in the summer when hydrological conditions are more favorable for bloom formation. Also, other organisms may provide conditions that are favorable for algal growth and the formation of HABs. In Lake Erie, for example, the introduction of zebra mussels in the late 1980s changed nutrient dynamics in the lake and thereby stimulated the growth of HABs, including *Microcystis* species. Apparently, zebra mussels are able to process phosphorus captured in the sediment and reintroduce it into the water column in a form that is accessible to cyanobacteria, including *Microcystis*. The loss of certain taxa may also promote HABs. Food web changes, such as the elimination or reduction in the population of grazers, may promote the formation of HABs.

Our understanding of the occurrence and impacts of HABs has expanded significantly in the last few decades, perhaps contributing to the greater identification of HABs in reservoirs and lakes. A number of historical accounts, however, suggest that HABs are not a recent phenomenon but instead have occurred for generations. One of the earliest suspected instances of an HAB occurred in 1931 in the Ohio River [7]. It is estimated that over 8000 people became ill following the ingestion of water contaminated by cyanobacteria from a tributary to the Ohio River. An epidemic in intestinal disorders in Charleston, West Virginia, was linked to the deterioration in water quality because of the presence of algae. Given the development and industrialization in the region at the time, wastewater discharges to the Ohio River most likely increased significantly in the years preceding the outbreak, thereby introducing conditions favorable for bloom formation.

In 1988, as a result of a cyanobacteria bloom in the Paulo Afonso region of Brazil, 88 people died and over 2000 became ill [8]. In one of the most noted HABs, the bloom occurred following the flooding of the newly constructed Itaparica Dam and reservoir and lasted for 42 days. Water samples were collected and analyzed and showed cyanobacteria levels of nearly 10,000 units/mL. Blood and fecal samples collected from patients with gastrointestinal illness pointed to cyanotoxins as the most probable cause of the outbreak.

Point-source discharges of nutrients from wastewater treatment plants were identified as a major cause of algal blooms in Lake Erie in the 1960s and 1970s. It is possible that some of the species of algae contributing to blooms in Lake Erie during this time period may have been harmful. The species *Microcystis*, for example, was identified during certain algal blooms on the lake during the 1960s and 1970s. Greater control of point sources through advanced wastewater treatment significantly reduced the occurrence of algal blooms in Lake Erie by the 1980s. In the 1990s, however, Lake Erie experienced a reemergence of algal blooms, blooms that have generally been more toxic in nature than the previous problems with algal blooms in the 1970s. HABs have remained a problem every year in Lake Erie since their reemergence in the late 1990s. A number of factors have been identified that may be contributing to the recent emergence of HABs on Lake Erie, such as changes in nonpoint sources of nutrients, the presence of zebra mussels, changing internal nutrient dynamics, changes in climate, and other factors.

Small, inland reservoirs are particularly vulnerable to HABs. In many cases, nonpoint sources of nutrients from large-scale agricultural production or other sources provide the necessary nutrients needed to support algal blooms, including HABs. These reservoirs are often shallow, allowing significant light penetration and warmer temperatures. Small reservoirs typically have low hydraulic residence times, which provide the time needed for algae to proliferate. As a result, ideal conditions are often present in many small, inland lakes and reservoirs to support the growth of algae. A significant HAB event occurred, for example, in Nebraska in 1988. Numerous people experienced serious gastrointestinal issues because of the ingestion of water tainted with harmful cyanobacteria. These conditions, and the increased awareness and understanding of HABs, greatly increased the identification of problematic outbreaks of harmful algae species. Compounding this problem is the fact that many of these small, inland reservoirs or lakes are also used as a source of drinking water for local communities. In some cases, small water systems have had to abandon a source of surface water as a result of HABs following the confirmation of the presence of toxic metabolites in the water after treatment.

With a greater understanding of the factors and conditions that promote HABs, the research community, lake managers, and government agencies are beginning to develop systems to predict HABs before they happen. In the Great Lakes, efforts are under way to provide seasonal forecasts as well as short-term (a few days) forecasts. The goal of these short-term forecasts is to provide government agencies, water utilities, and the public an early warning that the conditions exist to support an HAB. Other approaches for early warning systems have focused on monitoring toxin concentrations or proxies that may suggest toxins are present. As modeling capabilities and our understanding of HAB dynamics improve, better models and systems will likely emerge to predict the extent and magnitude of blooms in the future.

1.3 HABs and Drinking Water

The World Health Organization (WHO) and many countries worldwide have established enforceable limits or recommended guidelines for HAB toxins in drinking water. The WHO, for example, has established a guideline value for microcystin in drinking water of 1 µg/L. The United States has not yet established a maximum contaminant level for HAB toxins; however, this class of contaminant has been placed on the United States Environmental Protection Agency (USEPA) Contaminant Candidate List (CCL), demonstrating the need for further study for possible future regulatory action. The impact of a HAB bloom on drinking water will depend on a number of factors, including the magnitude and extent of the bloom, efforts to control bloom proliferation, the proximity of the bloom to the water treatment plant inlet, the conventional processes used for treatment and their particular operating parameters, as well as the availability of advanced treatment options at the treatment works. In a worst-case scenario, a bloom may occur adjacent to the inlet of the treatment works (as occurred in Toledo, Ohio in 2014), and the treatment plant may not be well prepared to provide optimal treatment to both minimize the release of toxins and reduce the concentration of specific toxins that may enter the plant or be released during conventional treatment operations. In a best-case scenario, only a fraction of the HAB cells or toxins produced enter the water treatment plant, and the facility has the treatment processes and experience to effectively deal with both the HAB cells and toxins.

Preventing the formation of HABs in freshwater lakes and reservoirs is preferable to dealing with the impacts to drinking and recreational waters during and after a bloom occurs. Changes to reduce nutrient inputs, however, may take years or even decades to implement. The identification of quantitative nutrient concentrations in a specific water body requires study and, often, years to establish. Even with suitable nutrient levels identified, controls on nonpoint sources and changes in land use practices can be difficult to implement. Also, lakes and reservoirs with a long history of nutrient input may have built up a significant reservoir of nutrients in the bottom sediments that can be reintroduced to the water column and stimulate HABs. The difficulty in reducing nutrient inputs to lakes and reservoirs, especially nonpoint sources of nutrients, has led to significant efforts, sometimes rather drastic, to control the occurrence and magnitude of HABs.

In Grand Lake–St. Marys, a small inland reservoir once part of the canal system in rural Ohio, HABs have had a major impact on local tourism and threaten drinking water. The lake is subject to a "perfect storm" of factors that contribute to the occurrence of HABs, including shallow depth, large nonpoint source inputs of nutrients, and extensive channelization of the watershed to support agricultural production. As a major component of the Ohio Canal system, Grand Lake–St. Marys is a man-made lake, constructed to provide equalization of water flow for canal operations. As a result, the lake is extremely shallow, with an average depth of less than 10 feet. Much of the area in the watershed feeding the lake is devoted to large-scale agricultural production of corn and soybeans, which results in significant nonpoint inputs of nitrogen and phosphorus to the lake. Finally, many of the streams in the watershed have been channelized to promote more effective drainage to prevent inundation of crops during heavy rains. As a result, the normal ecological processes that may reduce nutrient inputs to the lake are "short-circuited." This short circuiting occurs because of the shorter travel times of nutrients to the lake, which provide less opportunity for the uptake of nutrients by stream biota. The short circuiting also occurs because of the loss of ecological function caused by channelization, thereby reducing the "ecosystem services" provided by the stream—in this case, the service of nutrient processing.

In the face of these challenges, water managers have resorted to treating Grand Lake–St. Marys with aluminum sulfate, or "alum." Alum is a metal salt, long used for the coagulation and removal of turbidity in conventional drinking water treatment plants. When alum is added to water in a conventional water treatment plant, the aluminum ions react with hydroxide in the water to form a soluble aluminum hydroxide precipitate. This solid precipitate acts as a glue to facilitate the agglomeration of fine particles and enhance their settling in sedimentation basins. In waters containing phosphorus, aluminum ions may directly react with phosphate ions to form an insoluble aluminum phosphate precipitate, which can be subsequently removed by sedimentation. Or, under certain pH conditions, phosphate ions may adsorb to aluminum hydroxide precipitates and be removed during sedimentation as an adsorbed complex. To control HABs, alum is used to reduce phosphorus levels in the lake or reservoir through either precipitation or adsorption as described. Once associated with aluminum precipitates, the solids settle, and the phosphorus becomes "locked" in the lake sediments. The addition of alum during a bloom may also help in removing HAB cells from the water column.

Although the application of coagulants may help in reducing phosphate and HAB biomass, the application of such approaches is challenging. For one, application of chemicals like alum is costly for even a moderate-size lake. To minimize cost, a good understanding of the bloom location and extent is needed. If successful, it is still largely unknown what long-term impacts may occur as a result of the application of alum in natural systems. It is not known, for example, whether the phosphate locked within the sediments can be reintroduced into the water column if sediment chemistry changes or significant resuspension occurs. Also, the long-term ecological impacts of alum, especially for benthic species, are not well characterized.

Other approaches for eliminating, or at least minimizing, the frequency and magnitude of HABs are also being considered. In some locations, efforts are under way to mechanically destratify the water column to provide greater mixing and reducing the nutrient concentration at the surface of the water column. If alternative, low-nutrient stream flows are available, water managers may increase inputs to a lake or reservoir from these alternative sources to promote the flushing of nutrients from the system. Few surface water systems, however, have such opportunities. The addition of clay to a nutrient-impacted water body may act in a similar way as aluminum sulfate by promoting the adsorption and sedimentation of nutrients as well as aggregating biomass. For many decades, water managers have utilized various types of algicides to control blooms. In the case of HABs, however, the use of an algicide such as copper sulfate may disrupt the integrity of cell walls and result in significant release of HAB toxins. Last, biological methods have been explored to shift the ecosystem dynamics to inhibit HAB formation.

The last defense in protecting humans from HABs is to remove the toxins during drinking water treatment. To ensure removal of HABs from drinking water requires optimizing conventional treatment processes or introducing more advanced processes specifically tailored for removal or destruction of HAB toxins. Most conventional processes for treating drinking water, including coagulation, flocculation, sedimentation, and filtration, have little or no ability to remove HAB toxins once these toxins are introduced to the drinking water source. It is not surprising that these conventional processes have little ability to remove HAB toxins because their primary function is for the separation of solids and removal of fine particles and microorganisms. Conventional processes can be effective, however, in removing HAB cells. Care must be taken during processes such as coagulation and sedimentation to ensure vigorous mixing, filter backwashing, and solids pumping do not disrupt HAB cells and release intercellular toxins during the water treatment process.

Nontoxic algal metabolites, such as "taste-and-odor" compounds, have long been a problem for water utilities during algal blooms. As a result, many water treatment plants have implemented treatment processes or optimized existing processes to minimize these compounds in the finished water. The most common approach for removing taste-and-odor compounds, such as geosmin and methylisoborneol, is the application of either powdered activated carbon (PAC) or granular activated carbon (GAC). Both PAC and GAC have been shown to be effective at reducing taste-and-odor compounds and are commonly employed seasonally in many locations. As a result, significant effort has been carried out to examine whether PAC and GAC processes can also be used for the removal and control of HAB toxins. Although both taste-and-odor compounds and HAB toxins occur during algal blooms, these classes of contaminants have very different properties. Geosmin, on the one hand, is a relatively small molecule with a molecular weight of 182 g/mol; microcystin-LR is a roughly 1000 g/mol heptapeptide. Therefore, the optimal conditions for PAC or GAC treatment for the removal of geosmin may be different from the carbon needed for effective removal of microcystin or other HAB toxins. Although PAC or GAC can be implemented to reduce HAB toxins, these practices are not cheap and can significantly increase the cost of treatment. The cost to add PAC for a moderate-size city (e.g., 40 million gallons per day) may be upward of $1 million for the summer bloom season.

Chlorination, another typically employed process during conventional water treatment, may oxidize HAB toxins, thereby reducing their concentration in finished water. The effectiveness of chlorine, however, depends on the particular toxin as well as the form of chlorine (e.g., hypochlorous acid, chlorine dioxide, etc.). Although chlorine may be effective in destroying some HAB toxins, often large contact times are required, contact times not typically achieved in most water treatment plants. Also, chlorination may not completely destroy a particular toxin but instead transform the toxin into a by-product compound with unknown toxicity. As a strong oxidant, chlorine may disrupt any HAB cells able to circumvent previous treatment processes, such as coagulation, flocculation, sedimentation, or filtration. Even though nearly complete removal of HAB cells is expected in these separation processes under good operating conditions, failures in operation may provide the opportunity for the breakthrough of HAB cells to the chlorination process.

Ozonation has also been examined as a possible treatment process to destroy HAB toxins in drinking water. Like chlorination, ozonation is not effective for all classes of HAB toxins and can result in cell lysis, which may lead to the release of toxins into the water plant.

For particularly problematic situations, advanced treatment processes may be required. Membrane processes, such as reverse osmosis and "tight" nanofiltration, are effective at removing both HAB cells and HAB toxins, even toxins with the lowest molecular weight. While these membrane processes are effective, it can be difficult to apply them to the treatment of surface water because of excessive membrane fouling. Blooms of algae, and the associated organic matter, may deposit on membrane surfaces, resulting in a decrease in flux, increased pressure requirement, and shorter membrane lifetimes. Ultrafiltration (UF) may be effective at removing some HAB toxins if coupled to a sorption process such as with PAC. The PAC-UF process can provide efficient sorption and solids removal for HABs and HAB toxins but still suffers from the problems of membrane fouling.

Increasingly, advanced oxidation processes (AOPs) are being deployed to solve challenging drinking water quality problems. Even though AOPs show great promise, the demonstration of these processes at larger scales remains a challenge. Although ultraviolet (UV) radiation alone is generally not effective at destroying HAB toxins at the timescales needed for water treatment, advanced UV processes show promise. AOPs currently being explored are based on the photochemical destruction of HAB toxins in the presence of TiO_2 or peroxide. Additional research and development, however, are needed before these processes are viable for full-scale water treatment.

1.4 Overview

This introduction highlights what is currently known about HABs and HAB toxins, as well as some of the many challenges posed by HABs. This book explores many of these topics in greater detail, beginning with a discussion of the causes of HABs and their prevalence, both in the United States and globally. The chemical, hydrologic, and ecological factors controlling HAB occurrence, as well as the factors influencing the release of HAB toxins from HAB cells, are presented. Methods for reducing the occurrence of HABs depends on the development and application of policies and regulations; therefore, an overview of various policy approaches for reducing or minimizing HABs is discussed. When HABs occur, their impact can be mitigated through whole-lake approaches or improving water treatment processes. In the chapters that follow, approaches for reducing HAB formation and toxin release in lakes and reservoirs are outlined as well as the impact of conventional water treatment processes on HABs and HAB toxins. Last, new and more advanced treatment processes that offer promise for the future are presented.

The goal of this book is to provide a practical overview of the causes, impacts, and mitigation of HABs, with an emphasis on freshwater systems. Because many freshwater systems are used as a source of drinking water, options for minimizing or removing HAB toxins from the water we drink are highlighted. The book should be valuable to lake managers, engineers, researchers, policy makers, and others interested in reducing the impacts of HABs on ecosystems and humans.

References

1. C.B. Lopez, E.B. Jewett, Q. Dortch, B.T. Walton, and H.K. Hudnell, *Scientific Assessment of Freshwater Harmful Algal Blooms*. Interagency Working Group on Harmful Algal Blooms and Human Health of the Joint Subcommittee on Ocean Science and Technology., Washington, DC, 2008.

2. C. Svrcek and D.W. Smith, Cyanobacteria toxins and the current state of knowledge on water treatment options: a review. *Journal of Environmental Engineering and Science*, 3 (2004) 155–185.

3. H.K. Hudnell (Ed.), *Cyanobacterial Harmful Algal Blooms: State of the Science and Research Needs*, in: N. Back (Series Ed.), *Advances in Experimental Medicine and Biology*, Vol. 619. New York: Springer, 2008, p. 949.

4. H.K. Hudnell, The state of US freshwater harmful algal blooms assessments, policy and legislation. *Toxicon*, 55 (2010) 1024–1034.

5. A.M. Michalak, E.J. Anderson, D. Beletsky, S. Boland, N.S. Bosch, T.B. Bridgeman, J.D. Chaffin, K. Cho, R. Confesor, I. Daloğlu, J.V. DePinto, M.A. Evans, G.L. Fahnenstiel, L. He, J.C. Ho, L. Jenkins, T.H. Johengen, K.C. Kuo, E. LaPorte, X. Liu, M.R. McWilliams, M.R. Moore, D.J. Posselt, R.P. Richards, D. Scavia, A.L. Steiner, E. Verhamme, D.M. Wright, and M.A. Zagorski, Record-setting algal bloom in Lake Erie caused by agricultural and meteorological trends consistent with expected future conditions, *Proceedings of the National Academy of Sciences of the United States of America*, 110 (2013) 6448–6452.

6. D.J. Griffiths and M.L. Saker, The Palm Island mystery disease 20 years on: a review of research on the cyanotoxin cylindrospermopsin, *Environmental Toxicology*, 18 (2003) 78–93.

7. A.P. Miller, Epidemic of intestinal disorders in Charleston, W. Va., occurring simultaneously with unprecedented water supply conditions. *American Journal of Public Health*, 21 (1931) 198–200.

8. M.G. Teixeira, M.C. Costa, V.L. de Carvalho, M.S. Pereira, and E. Hage, Gastroenteritis epidemic in the area of the Itaparica Dam, Bahia, Brazil. *Bulletin of the Pan American Health Organization*, 27 (1993) 244–253.

2

Occurrence and Ecology of Harmful Algal Blooms

2.1 Introduction

This chapter begins by examining the types of harmful algal blooms (HABs) observed in freshwater systems. The term *harmful algal bloom* refers to a host of both cyanobacterial and noncyanobacterial species that create health problems for humans and animals. In 2008, the Interagency Working Group on Harmful Algal Blooms, Hypoxia, and Human Health developed the *Scientific Assessment of Freshwater Algal Blooms* [1], which reviewed current understanding of HAB occurrence in the United States. A discussion of the frequency of occurrence and the geographical distribution of HABs in freshwater systems is also presented in this chapter to establish the magnitude of the problem in the United States and globally. Once an understanding of the frequency and occurrence of HABs is developed, this chapter focuses on exploring the physical, chemical, and biological factors that are important in the formation of HABs and those factors that control the persistence and die-off of blooms. The ultimate impact of HABs on people and animals also depends on the expression and release of toxin from HAB species. This chapter summarizes the main factors that influence the "toxin quota" in HAB species, as well as the factors that influence toxin release in freshwater systems.

2.2 Types of Harmful Algal Blooms

Harmful algal blooms, or HABs, include a variety of organisms that create health issues for humans or animals or result in the deterioration of the aesthetic or recreational value of surface water [1]. Freshwater HABs can be divided into two broad classes: cyanobacterial and noncyanobacterial. Freshwater cyanobacteria typically classified as HABs include *Anabaena, Aphanizomenon, Aphanocapsa, Cylindrospermopsis, Hapalosphon, Lyngbya, Microcystis, Nodularia spumigena, Nostoc, Oscillatoria, Planktothrix,* and *Umezakia.* Many of the cyanobacterial HABs are filamentous, in that individual cyanobacterial cells affix to each other to form filaments. Filamentous forms include *Anabaena, Cylindrospermopsis, Lyngbya, Nodularia, Nostoc,* and *Planktothrix.* The filaments may be bead-like (e.g., *Anabaena*), flat cylinders (e.g., *Lyngbya*), or spiral in shape (e.g., the nontoxic *Spirulina*). Some cyanobacterial HABs, however, are not filamentous, including *Microcystis* and *Apanixomenon.* Both *Microcystis* and *Apanixomenon,* for example, form spherical colonies, and both can regulate buoyancy through various means. These filamentous and nonfilamentous cyanobacteria can produce a variety of toxins, such as hepatotoxins, neurotoxins, cytotoxins, and dermatotoxins, as well as other types.

A number of noncyanobacterial HABs are also known, including haptophytes, chlorophytes, macroalgae, euglenophytes, raphidophytes, dinoflagellates, cryptophytes, and diatoms. For example, important species of haptophytes contributing to HABs include *Prymnesium parvum* and *Chrysochromulina polylepis.* Ichthyotoxins, so-called fish toxins, are the primary toxins of concern produced by haptophytes. The euglenophyte *Euglena sanguinea* and various dinoglagellates are also known to produce ichthyotoxin. The other noncyanobacterial HABs mentioned do not produce known toxins but instead may lead to water discoloration, may form nuisance algal mats, or may contribute to hypoxia.

2.3 Occurrence of Freshwater HABs

Cyanobacterial HAB episodes have been identified in at least 35 states of the United States [1]. Although a comprehensive database is not available, the incidence of freshwater HABs is increasing in the United States [2, 3] and other parts of the world. In the United States, many of the largest lakes, including Lake Erie, Lake Michigan, Lake Champlain, Lake Ontario, Lake Huron, Lake Okeechobee, and Lake Poncharktrain, experience recurring blooms of harmful algae [4, 5]. In the western basin of Lake Erie, for example, blooms of *Microcystis* sp. have been documented since 1995 [5]. In 2011, a bloom of *Microcystis* on Lake Erie covered an area of roughly 3000 km^2 (see Figure 2.1), about twice the size of the island of O'ahu (see Chapter 1). The bloom of *Microcystis* on Lake Erie in 2011 was the largest recorded bloom on Lake Erie to date [6]. Blooms of *Microcystis* on Lake Erie are having a significant economic impact in the region, from reducing tourism to increasing the cost of purifying drinking water. In 2013, for example, Toledo, Ohio, spent an additional $1 million for treatment to combat HAB toxins in Lake Erie source water. Carroll Township, also on the Lake Erie shoreline, shut down operations in 2013 because of concerns about microcystin in Lake Erie water.

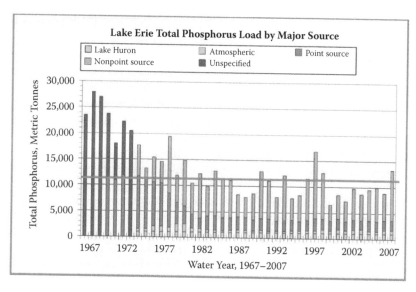

FIGURE 2.1 (See color insert.)

Annual loading of phosphorus to Lake Erie by major source. Reductions in point loads significantly reduced phosphorus inputs by the 1980s. (From the Ohio Lake Erie Phosphorus Task Force.)

Blooms of harmful algae are also prevalent in numerous smaller lakes, ponds, and reservoirs across the United States, including, but not exclusively in, Florida, Oregon, Nebraska, and Ohio [7]. In Ohio, for example, numerous cases of HABs have been documented, in addition to those in Lake Erie, including Grand Lake–St. Marys, Burr Oak State Park, Silver Lake, and others. In fact, HABs have become prevalent throughout the Midwest and Plains states, with documented HAB outbreaks in Indiana, Illinois, Kansas, Michigan, Minnesota, Wisconsin, Iowa, Texas, North Dakota, and South Dakota. In Nebraska, 26 health alerts and 69 health advisories were issued for lakes in the state in 2004 because of HABs.

In the western United States, HABs have been documented in Washington, Oregon, California, New Mexico, Nevada, Montana, Utah, Colorado, and Arizona. In Oregon, for example, HAB advisories have been issued for more than 25 lakes in the state. In the state of California, HABs have been documented in major rivers and lakes from Klamath River in the north to Lake Elsinore in the south.

In New York State, nearly 50% of samples collected from over 140 lakes tested positive for microcystin. Microcystin has been documented in numerous lakes across the state, such as Lake Ontario, Lake Champlain, Lake Neahtawanta, Oneida Lake, and dozens of other lakes from every geographical region.

In the Northeast, freshwater HABs have also been documented in the whole of New England (Vermont, Connecticut, Rhode Island, Maine, Massachusetts, and New Hampshire), as well as New Jersey and Pennsylvania.

The southeastern United States has been plagued with freshwater HABs, with every major state experiencing some type of outbreak. In Florida, cyanobacterial HABs have been known to occur since the 1970s, especially in Lake Okeechobee. In the late 1990s, Florida established the Harmful Algal Bloom Task Force and conducted a survey that found that over half of the water samples collected from over 75 water bodies tested positive for cyanotoxins. In Louisiana, Lake Pontchartrain experienced a significant bloom of *Anabaena, Microcystis,* and other HABs in 1997 as a result of the diversion of Mississippi River water into the lake. Although this HAB was a result of rather special hydrologic circumstances, it still demonstrates the widespread vulnerability of freshwater in this region.

Blooms of freshwater HABs are also well documented internationally, with known blooms on every major continent except Antarctica. A number of countries have experienced HABs with documented incidences of health-related problems, including Australia, Brazil, Canada, China, England, and Sweden [1]. For example, 52 dialysis patients are believed to have died as a result of exposure to cyanotoxins in Brazil in 1996. In Australia, gastrointestinal problems have been directly linked to the presence of cyanotoxins. The potential economic loss caused by freshwater HABs in Australia has been estimated in the hundreds of millions of dollars. In China, some studies suggested a connection between a higher incidence of liver cancer and drinking water contaminated with cyanotoxins.

2.4 Ecology of Harmful Algal Blooms

The occurrence of HABs is controlled by a variety of physical, chemical, and biological factors. Eutrophication is the driving force that leads to the formation of algal blooms. Eutrophication is the enrichment of a water body with certain nutrients, most typically nitrogen and phosphorus [8]. Although eutrophication promotes algae growth, the predominance of particular algal species is more nuanced. The factors that may influence the proliferation of a particular harmful algal species in a given aquatic system may be unique, with different freshwater HAB species responding to a different set of conditions. In Lake Erie, for example, eutrophication caused by phosphorus led to thick mats of cyanobacteria (*Anabaena, Aphanizomenon,* and *Microcystis*) and the filamentous green alga *Cladophora* [9] in the 1960s and 1970s. As a result, extensive research was carried out that ultimately led to limits on phosphorus inputs from point sources such as sewage treatment plants. The point source limits on phosphorus were successful in reducing phosphorus inputs to the lake. In fact, the corresponding reduction in phosphorus to the lake all but eliminated algal blooms by the 1980s. In the mid-1990s, however, significant blooms of cyanobacteria (*Microcystis* and *Lyngbya*) and green alga (*Cladophora*) began to appear in Lake Erie. The first massive bloom of *Microcystis* occurred in 2003, and similar blooms have occurred each year since.

To understand these changes in the distribution of algal species and magnitude of algal blooms, it is important to define the different operational forms of phosphorus. One measure of phosphorus in natural aquatic systems is *total phosphorus*, which represents the sum of dissolved phosphorus and particulate phosphorus. In conducting a water quality analysis, particulates (including particulate phosphorus) are defined as the material retained on a 0.45-μm filter. The dissolved constituents are those that pass through the 0.45-μm filter and operationally may also include colloidal material. Dissolved reactive phosphorus (DRP) is a component of dissolved phosphorus that readily reacts with molybdate and is considered 100% bioavailable. Particulate phosphorus is generally considered only 30% bioavailable.

The "reeutrophication" of Lake Erie underscores the complex relationship between physical, chemical, and biological factors that controls the proliferation of HABs in most aquatic systems. In Lake Erie, a number of theories have been proposed to explain the reeutrophication and shift in species distribution with the predominance of HABs. One possible explanation is greater internal loading of phosphorus mediated by the processing of phosphorus in sediment as a result of the introduction of invasive zebra mussels or quagga mussels. The zebra and quagga mussels, which also were first observed in Lake Erie in the 1980s, may be able to release this previously unavailable form of phosphorus "locked away" in the sediments. This explanation is consistent with the observed increase in open-water total phosphorus observed in Lake Erie in more recent years, despite relatively constant total phosphorus loading to the lake. In addition to recycling phosphorus into the water column, mussels reduce particulates through filter feeding, which increases water clarity and promotes algae growth deeper into the water column.

Although total phosphorus loads to Lake Erie have remained relatively constant during this reeutrophication period (see Figure 2.1), water quality sampling revealed an increased loading of DRP [9], which is nearly 100% available to algae. In the 1970s, less than 20% of the total phosphorus loading to Lake Erie from the Maumee and Sandusky watersheds was in the form of DRP. During the 1980s, loads of DRP to Lake Erie decreased as total phosphorus concentrations declined. Starting in the mid-1990s, however, the loading of bioavailable phosphorus to Lake Erie from the Maumee and Sandusky watersheds began to increase. In fact, loading of DRP from these watersheds is higher now than at any time since 1975. It has been suggested that DRP loading to Lake Erie from the Maumee and Sandusky watersheds is significantly affected by a decrease in the capacity of streams to assimilate phosphorus. Decreased retention time of nutrients in streams discharging to Lake Erie reduces the potential for nutrient assimilation prior to reaching the lake.

Other factors, including climate change and urban storm water, have also been suggested as possible contributing factors to the emergence of HABs on Lake Erie in recent decades. Climate change-induced impacts on water temperature, wind, and lake water levels may be having ecosystem-level impacts that increase the chances for HABs. For example, lower lake levels and higher water temperature generally promote algae growth. Changes in the direction and speed of wind on Lake Erie might influence stratification. Thermal stratification, for example, is characterized as a change in water temperature as a function of depth. Stratification has a significant impact on the amount of mixing that may occur between different water depths. Increasing wind speeds, for example, induce greater turbulence into the water column, which may destratify a water body and lead to greater mixing. Therefore, the extent of stratification will affect water temperature (as a function of depth) and transport of nutrients from the sediment to the water surface, both of which will have an impact on the algae growth. Urban storm water runoff may also be a significant source of bioavailable phosphorus to Lake Erie as a result of suburban fertilizer use, stream bed erosion, and runoff from commercial and industrial properties.

By taking a closer look at HABs in Lake Erie and other aquatic systems, a number of factors can be identified that influence the extent and characteristics of HABs. Factors such as temperature, nutrient loading (magnitude and characteristics) and cycling, water clarity, lake hydrodynamics, hydrology, climate and weather, and biological community structure all can have an impact on severity and characteristics of an HAB. The impact of these various factors can be illustrated by looking at the typical annual cycle of *Microcystis*. Blooms of *Microcystis* in winter and spring are uncommon because cyanobacterial cells tend to "winter" in lake and tributary sediment. During the spring and summer, *Microcystis* will recolonize the surface water through bioturbation and turbulent mixing of the sediments. Once the water temperature increases above about 15°C, *Microcystis* may begin to proliferate. During the fall, water temperatures decline, and the potential for blooms of *Microcystis* decline.

Eutrophication and the availability of nutrients are the primary drivers for the formation of HABs. In most natural aquatic systems, either nitrogen or phosphorus is the limiting nutrient. Phosphorus is generally the limiting nutrient in temperate, freshwater lakes, and nitrogen is limiting in coastal bays in estuaries [8]. Research suggested that a major factor for this observation is the greater availability of phosphorus in coastal and estuarial systems as a result of the sequestration by iron sulfide. The salinity of seawater provides greater sulfur for the formation of iron sulfide solids, which limits the iron available for the sequestration of phosphorus. Other factors, of course, may also play a role in phosphorus limitations in lakes, such as lower nitrogen fixation by cyanobacteria and higher overall rates of denitrification in coastal waters, differences in the phosphorus requirements of biota in freshwater and marine systems, and differences in nutrient processing and dynamics in these different systems.

The absolute amount of nitrogen or phosphorus, as well as the relative ratio of nitrogen to phosphorus, is important in understanding HABs. As suggested by data from Lake Erie and other natural aquatic systems, blooms of *Microcystis* tend to correlate with increasing total phosphorus concentrations. Previous studies suggest that at lower phosphorus concentrations (<10 µg/L), green algae outcompete *Microcystis* for available phosphorus [10]. Above this threshold, sufficient phosphorus is available to support blooms of *Microcystis*. At total phosphorus concentrations of 100 µg/L, the probability that a bloom will form is at a maximum of about 80%. Above this concentration, increasing inputs of phosphorus do not appear to further stimulate the potential for harmful blooms of *Microcystis*.

The impact of nitrogen on HABs is dependent on the nitrogen-fixing ability of algae and cyanobacteria. Some algae and cyanobacteria (e.g., *Anabaena* and *Aphanizomenon*) are able to directly fix nitrogen from the atmosphere, but others (e.g., *Microcystis*) are not. During conditions of nitrogen limitation, therefore, species that are able to fix nitrogen are at a competitive advantage compared to species that cannot directly fix nitrogen. Therefore, blooms of *Microcystis* may tend to die away as nitrogen becomes limiting. This observation may explain the greater prevalence of blooms of *Microcystis* in Lake Erie in recent decades compared to the 1970s. During the 1970s, greater levels of total phosphorus resulted in large blooms of algae that then subsequently depleted available nitrogen. As a result, nitrogen-fixing species, such as *Anabaena* and *Aphanizomenon* were favored, which is consistent with the known composition of algal blooms at that time. In the 1990s at least, controls on point sources of phosphorus reduced the amount of total phosphorus discharged into the lake, which may have reduced the potential for nitrogen limitation, thus providing conditions more amenable to blooms of non-nitrogen-fixing species like *Microcystis*. The form of nitrogen can also play a role in controlling the extent and magnitude of an HAB. *Microcystis*, for example, prefers ammonia nitrogen over nitrate.

The N:P ratio is an important factor to consider in any nutrient management strategy aimed at reducing the potential for HABs. Nutrient management strategies that focus on reducing phosphorus loading, for example, will tend to increase the N:P ratio because, although phosphorus levels decline, levels of nitrogen stay constant. As a general rule (see Figure 2.2), in aquatic systems characterized by a low molar ratio of nitrogen to phosphorus (<15), cyanobacteria will tend to dominate [3, 11]. At higher molar ratios (>20), on the other hand, algal species will be at a competitive advantage. The predominance of cyanobacteria at low N:P ratios is attributed to the nitrogen-fixing ability of many blue-green algae, which allows them to flourish in

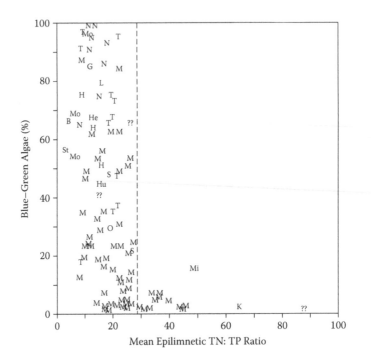

FIGURE 2.2
Relation between N:P ratio and predominance of blue-green algae. (Reprinted from V.H. Smith, *Science*, 221 (1983) 669–670.)

competition with other species not able to fix nitrogen. At higher N:P ratios, nitrogen-fixing cyanobacteria are less effective at competing for phosphorus compared to other phytoplankton. *Microcystis*, of course, is not able to fix nitrogen and therefore would not fit this general rule of thumb. As discussed, relatively low N:P ratios in Lake Erie in the 1970s were a result of the higher phosphorus loading at the time and potentially led to nitrogen-limiting conditions that favored nitrogen-fixing species rather than *Microcystis*. Various other factors make the applicability of a general rule of thumb problematic given the complex relationship between HABs and turbidity, hydrodynamics, and light levels.

Although most states are focusing on establishing phosphorus limits for freshwater systems, current research suggests the control of both nitrogen and phosphorus may be required in some nutrient-impacted water bodies to bring about significant water quality improvements [12]. Controlling phosphorus is especially important for nitrogen-fixing HAB species such as *Anabaena*, *Aphanizomenon*, *Cylindrospermopsis*, and *Nodularia*. For HAB species not able to fix nitrogen, such as *Microcystis*, *Lyngbya*, and *Planktothrix*, the control of both nitrogen and phosphorus is important.

2.5 Toxin Release and Modes of Expression

Although the extent of an HAB, in terms of biomass, is important, it is also critical to consider how various factors affect the toxicity of HABs and toxin release. Not all species of *Microcystis*, for example, are toxic during a bloom. Also, the expression of toxin by *Microcystis* may change with time under different conditions. The toxicity of a particular species of *Microcystis* depends on the presence of the mycA-J gene cluster and expression of associated proteins. The mycA-J gene cluster codes for the production of a peptide synthetase and polyketide enzyme complex that are involved in the production of microcystin. Gene clusters have also been identified for the synthesis of saxitoxin (i.e., *stxA*, *stxB*, etc.), cylindrospermopsin (i.e., *cyrA*, *cyrB*, etc.), and anatoxin (i.e., *anaA*, *anaB*, etc.).

The concentration of toxic metabolites generated during an HAB will depend on the quantity of toxic algal or cyanobacterial species as well as the "toxin quota," or mass concentration of toxin relative to the species cell mass. Field studies in Lake Taihu, China, for example, showed that both *Microcystis* cell density and microcystin cell quota correlated positively with lake microcystin levels [13]. Thus, a bloom of low biomass but high toxin quota may be especially serious even if visual observation of the bloom is less dramatic. On the other hand, blooms characterized by a high amount of biomass but small toxin quota may be less serious than expected based on visual observation alone. One of the earliest factors investigated was cell growth rate, with some studies suggesting a positive correlation between microcystin quota and cell growth rate [14]. However, later studies have contradicted this earlier finding [15], so the role of growth rate on toxin cell quota is still somewhat unclear.

Studies have suggested a connection between nutrient levels and toxin quota, unrelated to the effect of nutrients on growth rate. In laboratory culture experiments, for example, Downing et al. [16] showed a positive correlation between nitrate concentration and microcystin cell quota. Lee et al. [17] and Horst et al. [18] came to a similar conclusion regarding the effect of nitrogen on microcystin quota. In Lake Erie, laboratory and field data both support the idea that microcystin quota increases with increasing nitrogen availability. The increase in microcystin quota in Lake Erie was attributed to phenotypic changes occurring in the *Microcystis* plankton community [18] and were not simply the result of changing growth rates in response to higher nitrogen levels.

Min et al. [13] carried out a statistical analysis of field-sampling data from Lake Taihu, China, and determined that microcystin cell quotas were most dependent on water conductivity, dissolved inorganic carbon (DIC), and water temperature. Microcystin cell quotas increased, for example, as a result of decreasing water conductivity. It was speculated that conductivity provides a bulk measure of the availability of biologically important elements, with decreasing conductivity leading to shortages of critical elements such as iron. It was speculated that under limiting conditions (e.g., limited iron concentrations), *Microcystis* produces toxic metabolites as a means to out-compete other organisms for limited elemental resources. In a similar way, decreasing DIC also led to greater microcystin cell quotas. Under conditions of low DIC, the greater production of microcystin provides *Microcystis* with a competitive advantage in photosynthesis over other phytoplankton. The growth rate of *Microcystis* generally increases with temperature up to about 32°C, with declining growth at higher temperatures [19].

2.6 Models and Early Warning Tools

An understanding of the factors and processes influencing HABs provides an opportunity to develop models to better understand and predict future blooms and bloom dynamics. A number of different types of models have been developed over the years to understand HABs (for an early review of such models, see [20]). Early, simple models of HABs consisted of aggregated biological models with little or no physical factors. For example, early models for predicting the proliferation of "red tide" were based on the relative rates of grazing and nutrient uptake. Slightly more sophisticated models have incorporated different biological species into the system but still lacked a detailed description of the physics. Current models and research in this area are based on coupled systems of equations describing the physical-biological systems with varying complexity.

Detailed physical-biological models are being combined with satellite data, field measurements, and buoy data to forecast possible HABs and provide for early warning of the public and government agencies. In marine systems, the National Oceanic and Atmospheric Administration (NOAA) Harmful Algal Bloom Operational Forecast System (HAB-OFS) aims to mitigate the impacts of HABs through early detection and forecasting. Forecasts of *Karenia brevis* blooms (so-called red tide) are developed twice a week during the most active season and posted on the NOAA website. The forecasts predict the potential severity of blooms of *Karenia brevis* over the next 3–4 days. A similar forecasting system has been developed for Lake Erie to predict blooms of *Microcystis* and is updated once per week [21, 22]. An example bulletin is shown in Figure 2.3.

The forecasting system for HABs in Lake Erie incorporates over 11 years of nutrient data, other field data, and the medium spectral resolution imaging spectrometer (MERIS) on the European Space Agency satellite, Envisat. The satellite data are used to determine a "cyanobacteria index" (CI) for past conditions. Wind speed and water temperature data are collected from buoys on Lake Erie maintained by the National Data Buoy Center (NDBC). The movement of the bloom is predicted using a hydrodynamic circulation model, based on the Great Lakes Forecasting System (GLFS) [23]. The GLFS is a hydrodynamic model based on the Princeton Ocean Model (POM). The POM is a free-surface ocean model and incorporates the effects of turbulence. The GLFS was originally created to generate nowcasts and forecasts of currents, water levels, waves, and temperature in the Great Lakes, including Lake Erie. The application of the GLFS for Lake Erie enables the prediction of bloom movement based on predicted lake currents. Vertical mixing and cyanobacteria reproduction are not yet included in the model. Given the buoyancy of *Microcystis* and the shallowness of the western basin of Lake Erie, the experimental HAB forecasting tool for Lake Erie assumes that vertical mixing is not significant. Predictions are made for a few days in the future. Over such timescales, cyanobacteria reproduction is not expected to significantly alter the forecasts.

More biologically sophisticated, multiparameter physiological models are beginning to be developed to capture the life stages and trophic interactions of HAB species (for a review, see [24]). For example, Hood et al. [25] developed a semi-idealized ecosystem model to predict the population dynamics of *Pfiesteria*. The dynamic systems model links the population dynamics of *Pfiesteria* to the three main population drivers: nutrients, predation, and turbulence. In so doing, the model attempts to predict the various life stages of *Pfiesteria*, including flagellated, amoeboid, and encysted, under different conditions, as well as the predominance of toxic versus nontoxic strains. Application of the model confirmed field observations that toxic strains of *Pfiesteria* are promoted when grazing is inhibited and that nontoxic strains predominate in more turbulent, nutrient-rich environments.

Ultimately, researchers would like to link models predicting bloom dynamics to social-economic models to develop better policies to prevent HABs. Lake Erie researchers, for example, have attempted to link physical hydrodynamic models, ecosystem models, and models for understanding social interactions and economics. The physical hydrodynamic models help understand water currents, waves, and temperature and how these parameters affect lake chemistry and biological and other functions. Ecological models were designed to better understand the relationship between invasive mussels, phosphorus, and HABs. Last, the hydrodynamic and ecological models were coupled to social system models to link the physics and ecology of Lake Erie to specific management options. The goal of the modeling was to determine the optimal economic conditions and management policies that both support the economy as well as the Lake Erie ecosystem. The Lake Erie system is essentially governed by two competing interests: the farm industry, which relies on nutrient inputs for economically viable production, and the tourist industry, which requires pristine lake conditions to entice visitors and preserves the ecosystem services provided by Lake Erie. The modeling uncovered two stable "states" in the Lake Erie physical-ecological-social

Experimental Lake Erie Harmful Algal Bloom Bulletin

National Centers for Coastal Ocean Science and Great Lakes Environmental Research Laboratory

17 September 2012; Bulletin 16

The bloom has weakened in Maumee Bay since last week's bulletin. The water temperature is beginning to cool which could further decrease growth. Last Thursday, a Public Health Advisory was posted for Maumee Bay State Park beach. We forecast for a S-SE transport.

- Dupuy, Wynne, Briggs, Stumpf

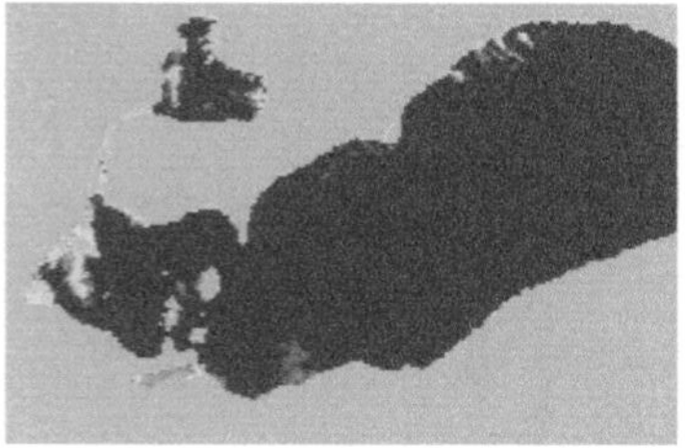

Figure 1. MODIS Cyanobacterial Index from 16 September 2012.

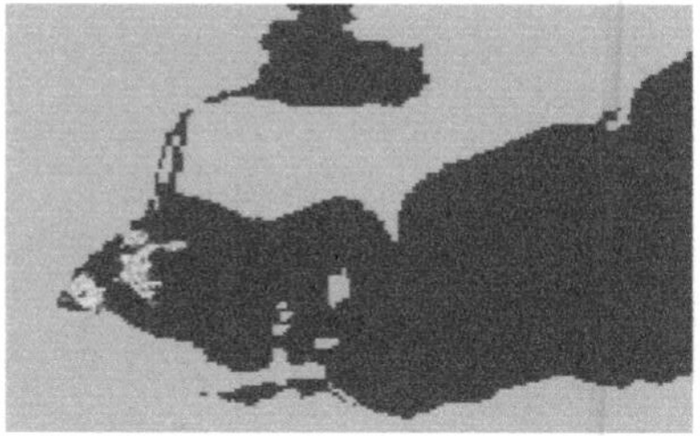

Figure 2. Nowcast position of bloom for 17 September 2012 using GLCFS modeled currents to move the bloom from the 16 September 2012 image.

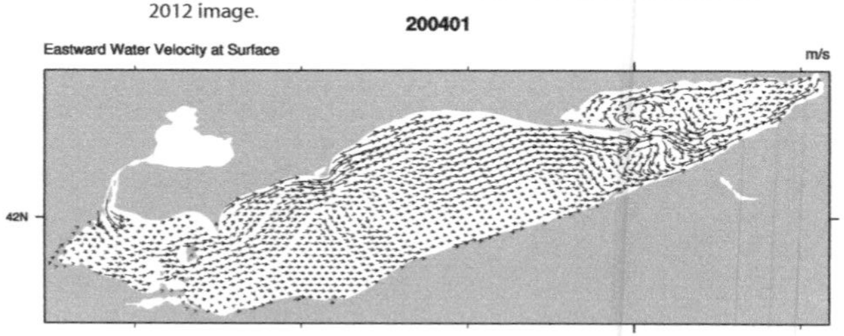

Averaged forecasted currents from Great Lakes Coastal Forecasting System over the next 72 hours.

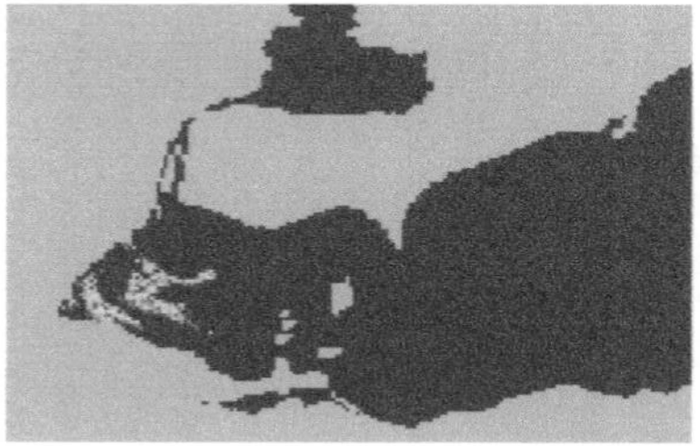

Figure 3. Forecast position of bloom for 20 September 2012 using GLCFS modeled currents to move the bloom from the 16 September 2012 image.

To subscribe to this bulletin, go to :
http://www.glerl.noaa.gov/res/Centers/HABS/lake_erie_hab/signup.php

FIGURE 2.3 (See color insert.)
Example of NOAA HAB bulletin for Lake Erie. *(continued)*

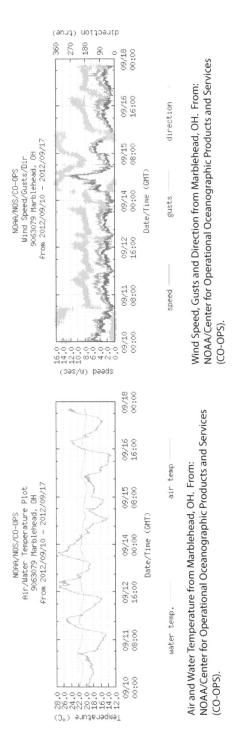

Wind Speed, Gusts and Direction from Marblehead, OH. From: NOAA/Center for Operational Oceanographic Products and Services (CO-OPS).

Air and Water Temperature from Marblehead, OH. From: NOAA/Center for Operational Oceanographic Products and Services (CO-OPS).

To subscribe to this bulletin, go to:
http://www.glerl.noaa.gov/res/Centers/HABS/lake_erie_hab/signup.php

FIGURE 2.3 (continued) (See color insert.)
Example of NOAA HAB bulletin for Lake Erie.

system: (1) a pristine state with few people and little economic activity and as a result greatly curtailed phosphorus inputs to the lake and (2) a state that not only supports a moderate amount of economic activity (and people) but also entails a moderate amount of ecological degradation because of phosphorus.

A related effort by Roy et al. [26] developed an integrated ecological-social model to explore how societal preferences have an impact on nutrient management strategies. The integrated model consisted essentially of an ecological model for the Sandusky Bay of Lake Erie and a phosphorus management model (see Figure 2.4). The ecological model for Sandusky Bay incorporated physical and chemical variables such as water flow, temperature, clarity, and phosphorus loading, as well as biological components to represent invasive mussels, zooplankton, and algal dynamics. The phosphorus management model attempted to incorporate societal preferences, stakeholder (e.g., environmental groups, phosphorus management groups) pressure, and benefits to farmers and the environment. The research suggested that a systemic approach is needed that involves a combination of upstream phosphorus load reductions, ecosystem restoration, and investment in low-cost nutrient efficiency technology. The modeling exercise highlighted the idea that eutrophication is as much social, cultural, and political as it is ecological.

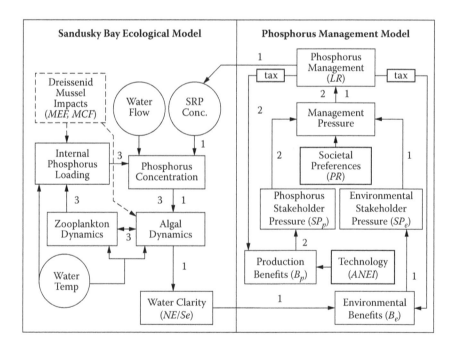

FIGURE 2.4
Integrated ecological-social model for Lake Erie. (From E.D. Roy, J.F. Martin, E.G. Irwin, J.D. Conroy, and D.A. Culver, *Ecological Engineering*, 37 (2011) 1661–1672.)

References

1. C.B. Lopez, E.B. Jewett, Q. Dortch, B.T. Walton, and H.K. Hudnell, *Scientific Assessment of Freshwater Harmful Algal Blooms*. Interagency Working Group on Harmful Algal Blooms and Human Health of the Joint Subcommittee on Ocean Science and Technology, Washington, DC, 2008.
2. H.K. Hudnell, The state of US freshwater harmful algal blooms assessments, policy and legislation. *Toxicon*, 55 (2010) 1024–1034.
3. H.K. Hudnell (Ed.), *Cyanobacterial Harmful Algal Blooms: State of the Science and Research Needs*, in: N. Back (Series Ed.), *Advances in Experimental Medicine and Biology*, Vol. 619. New York: Springer, 2008, p. 949.
4. H.W. Pearl, N.S. Hall, and E.S. Calandrino, Controlling harmful cyanobacterial blooms in a world experiencing anthropogenic and climatic-induced change, *Science of the Total Environment*, 409 (2011) 1739–1745.
5. G.L. Boyer, *Monitoring Harmful Algal Blooms in the Great Lakes*. Great Lakes Research Consortium, Syracuse, NY, 2006, p. 34.
6. R.P. Stumpf, T.T. Wynne, D.B. Baker, and G.L. Fahnenstiel, Interannual variability of cyanobacterial blooms in Lake Erie. *PLOS ONE*, 7 (2012) e42444.
7. M. Burch, Effective doses, guidelines and regulations, in: K.H. Hudnell (Ed.), *Cyanobacterial Harmful Algal Blooms: State of the Science and Research Needs*. New York: Springer, 2008, p. 833.
8. A.J. Horne and C.R. Goldman, *Limnology*. New York: McGraw-Hill, 1994.
9. OLEPT Force, OEP Agency, *Ohio Lake Erie Phosphorus Task Force Final Report*. Ohio Environmental Protection Agency, Columbus, 2010.
10. C.E.W. Steinberg and H.M. Hartmann, Planktonic bloom-forming *Cyanobacteria* and the eutrophication of lakes and rivers. *Freshwater Biology*, 20 (1988) 279–287.
11. V.H. Smith, Low nitrogen to phosphorus ratios favor dominance by blue-green algae in lake phytoplankton. *Science*, 221 (1983) 669–670.
12. D.J. Conley, H.W. Pearl, R.W. Howarth, D.F. Boesch, S.P. Seitzinger, K.E. Havens, C. Lancelot, and G.E. Likens, Controlling eutrophication: nitrogen and phosphorus. *Science*, 323 (2009) 1014–1015.
13. T. Min, X. Ping, C. Jun, Q. Boqiang, Z. Dawen, N. Yuan, Z. Meng, W. Qing, and W. Laiyan, Use of a generalized additive model to investigate key abiotic factors affecting microcystin cellular quotas in heavy bloom areas of Lake Taihu. *PLOS ONE*, 7 (2012) e32020. doi:10.1371/journal.pone.0032020.
14. B.M. Long, G.J. Jones, and P.T. Orr, Cellular microcystin content in N-limited *Microcystis aeruginosa* can be predicted from growth rate. *Applied and Environmental Microbiology*, 67 (2000) 278–283.
15. S. Jähnichen, T. Ihle, and T. Petzoldt, Variability of microcystin cell quota: a small model explains dynamics and equilibria. *Limnologica—Ecology and Management of Inland Waters*, 38 (2008) 339–349.
16. T.G. Downing, C. Meyer, M.M. Gehringer, and M. van de Venter, Microcystin content of *Microcystis aeruginosa* is modulated by nitrogen uptake rate relative to specific growth rate or carbon fixation rate. *Environmental Toxicology*, 20 (2005) 257–262.
17. S.J. Lee, M.H. Jang, H.S. Kim, B.D. Yoon, and H.M. Oh, Variation of microcystin content of *Microcystis aeruginosa* relative to medium N:P ratio and growth stage. *Journal of Applied Microbiology*, 89 (2000) 323–329.

18. G.P. Horst, O. Sarnelle, J.D. White, S.K. Hamilton, R.B. Kaul, and J.D. Bressie, Nitrogen availability increases the toxin quota of a harmful cyanobacterium, *Microcystis aeruginosa*. *Water Research*, 54 (2014), 188–198.

19. C. Nalewajko and T.P. Murphy, Effects of temperature, and availability of nitrogen and phosphorus on the abundance of *Anabaena* and *Microcystis* in Lake Biwa, Japan: an experimental approach. *Limnology*, 2 (2001) 45–48.

20. P.J.S. Franks, The ecology and oceanography of harmful algal blooms. *Limnology and Oceanography*, 42 (1997) 1273–1282.

21. T.T. Wynne, R.P. Stumpf, M.C. Tomlinson, G.L. Fahnenstiel, J. Dyble, D.J. Schwab, and S.J. Joshi, Evolution of a cyanobacterial bloom forecast system in western Lake Erie: development and initial evaluation. *Journal of Great Lakes Research*, 39, Supplement 1 (2013) 90–99.

22. T.T. Wynne, R.P. Stumpf, M.C. Tomlinson, D.J. Schwab, G.Y. Watabayashi, and J.D. Christensen, Estimating cyanobacterial bloom transport by coupling remotely sensed imagery and a hydrodynamic model. *Ecological Applications*, 21 (2011) 2709–2721.

23. D.J. Schwab and K.W. Bedford, The Great Lakes forecasting system. *Coastal and Estuarine Studies*, 56 (1999) 157–173.

24. J. Heisler, P.M. Glibert, J.M. Burkholder, D.M. Anderson, W. Cochlan, W.C. Dennison, Q. Dortch, C.J. Gobler, C.A. Heil, E. Humphries, A. Lewitus, R. Magnien, H.G. Marshall, K. Sellner, D.A. Stockwell, D.K. Stoecker, and M. Suddleson, Eutrophication and harmful algal blooms: a scientific consensus. *Harmful Algae*, 8 (2008) 3–13.

25. R.R. Hood, X. Zhang, P.M. Glibert, M.R. Roman, and D.K. Stoecker, Modeling the influence of nutrients, turbulence and grazing on *Pfiesteria* population dynamics, *Harmful Algae*, 5 (2006) 459–479.

26. E.D. Roy, J.F. Martin, E.G. Irwin, J.D. Conroy, and D.A. Culver, Living within dynamic social-ecological freshwater systems: System parameters and the role of ecological engineering. *Ecological Engineering*, 37 (2011) 1661–1672.

3

Toxin Properties, Toxicity, and Health Effects

3.1 Introduction

Harmful algal blooms (HABs) and their associated toxins are a significant threat to human health. Cyanobacteria and other HAB organisms produce a variety of toxins, including neurotoxins, hepatotoxins, dermatotoxins, and other bioactive compounds. As discussed in Chapter 1, the most typically cited classes of HAB toxins include microcystins, saxitoxins (STXs), anatoxins, and cylindrospermopsin. Although complete understanding of the health effects of HABs and HAB toxins has not yet been developed, HAB toxins are known to cause a variety of short- and long-term health effects in animals and humans. In natural systems, HABs result in fish kills and may lead to death or acute health effects in animals and livestock. Short-term acute effects of HAB toxins on humans include rashes, liver inflammation, numbness, dermatitis, gastrointestinal problems, liver failure, and others. Documented longer-term impacts associated with low-level chronic exposure to HAB toxins are less well known but potentially include tumor formation, cardiac arrhythmia, and liver failure.

From a risk assessment perspective, it is important to identify the various factors and processes that define the risk associated with contact to HABs and HAB toxins, in terms of both human health and the environment. As a general framework, it is important to identify sensitive populations, elucidate the fate and transport of HAB cells or toxins in natural or engineered systems, understand the significance of different routes of exposure, and ultimately understand the toxicity of the compounds or other health effects on contact. Many of the known cases of poisoning by HABs or HAB toxins occurred in populations of more sensitive individuals, such as children or people undergoing dialysis. During a bloom, the HAB cells or toxins often must be transported in the water column to come into contact with humans or animals. This transport may involve the movement of HAB cells by wind or currents, the diffusion or advection of HAB toxins, or a variety of other transport processes. HAB toxins may also be degraded in the natural environment through biological interactions, sunlight, adsorption, or other processes. On contact with humans or animals, the route of exposure (dermal, ingestion, inhalation) and the acute, short-term, or chronic toxicity of the compounds controls the ultimate impact on human or animal health.

This chapter describes the physical and chemical properties of algal toxins, including microcystins, anatoxin, nodularin, cylindrospermopsin, and STX. The focus or goal of this chapter is to describe the properties of these compounds important in controlling their fate, transport, and toxicity. This chapter reviews what is known about the various modes of toxicity of HAB toxins, such as acute, short-term, and chronic toxicity; genotoxicity; and the potential to cause cancer. HABs may produce a variety of toxins, such as hepatotoxins, neurotoxins, cytotoxins, and dermatotoxins. The United States Environmental Protection Agency (USEPA) has reviewed the toxicology research for microcystin, cylindrospermopsin, and anatoxin-a as part of the chemical review process for promulgation of new drinking water standards [1, 2]. Processes controlling the fate and transport of HAB toxins are discussed in a subsequent chapter.

Humans and animals may come into contact with HABs and HAB toxins through a variety of routes. A major exposure risk is via recreational waters. The exposure of both humans and animals to HABs and HAB toxins through recreational waters has been established as one of the most common exposure routes. Dermal exposure, ingestion, and inhalation may all occur during contact with recreational waters. There have also been documented cases of exposure to HAB toxins via drinking water and during dialysis treatment. Last, an important route of exposure is through the ingestion of seafood or dietary supplements synthesized from algae.

The following sections of this chapter describe the most significant health effects of HAB toxins, with an emphasis on the available epidemiological cases. The chapter begins with the discussion of nodularin, as this was one of the first suspected HAB toxins, with follow-up on the properties, toxicity, and health effects of other important classes of HAB toxins.

3.2 Nodularin

In a letter to the editor of *Nature* in 1878, George Francis made the first connection between algae and harmful effects to animals [3]. In his letter, he described *Nodularia spumigera* "forming a thick scum like green oil paint, some two to six inches thick" in Lake Alexandrina, which sits at the mouth of the Murray River, near the city of Adelaide in southern Australia. Apparently, it had been a warm summer in 1877, resulting in low-flow conditions and elevated water temperatures (76°F at the surface) in Lake Alexandrina. Francis noted his visual observation of livestock drinking from the lake and detailed the effects of drinking the contaminated water on cattle, sheep, pigs, dogs, and horses. He described how the contaminated water "acts poisonously, and rapidly causes death; symptoms—stupor and unconsciousness, falling and remaining quiet, as if asleep, unless touched, when convulsions come on, with head and neck drawn back by rigid spasm, which subsides before death." He detailed the time from ingestion to the onset of symptoms and ultimate death for the different animals.

Lake Alexandrina continues to experience algal blooms today (see Figure 3.1). In fact, *Nodularia* has been identified in Lake Alexandrina in recent studies [4]. The continued presence of *Nodularia* in Lake Alexandrina is of greater concern in the present because of the placement of a number of dams on the lake to reduce saltwater. In fact, the total dissolved solids level of the lake is low, and as a result, water from the lake is used by a number of drinking water utilities [4]. In 1991, over 100 years after Francis first described the bloom of *Nodularia* in Lake Alexandrina, eight individuals became ill because of contact with *Nodularia* on the Murray River [5, 6]. Symptoms included skin and eye irritation as well as respiratory symptoms. Blooms of *Nodularia* have been found in numerous lakes in Australia, the Baltic Sea, German North Sea, North America, and other parts of the world.

FIGURE 3.1 (See color insert.)
Algal bloom in Lake Alexandrina in southern Australia.

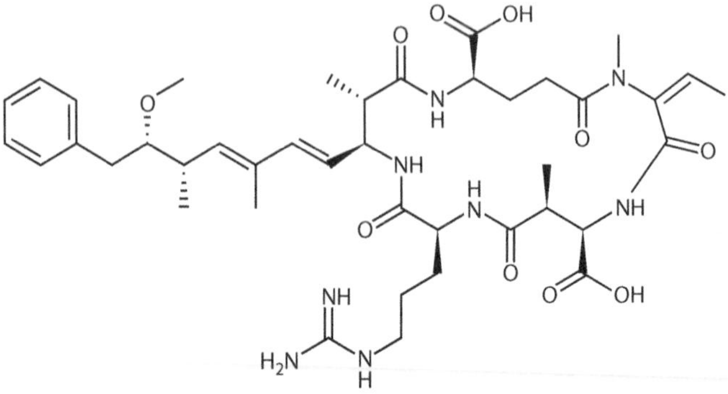

FIGURE 3.2
Primary structure of nodularin.

3.2.1 Nodularin Properties

The chemical formula for nodularin is $C_{41}H_{60}N_8O_{10}$, and the toxin has a molecular weight of 824.96. The compound is therefore one of the larger HAB toxins, just somewhat smaller than the microcystins. The chemical structure of nodularin is shown in Figure 3.2. Nodularin is a cyclic pentapeptide containing the unique β-amino acid ADDA ((2S,3S,4E,6E,8S,9S)-3-Amino-9-methoxy-2,6,8-trimethyl-10-phenyldeca-4,6-dienoic acid), which is also found in microcystin, D-glutamic acid, N-methyldehydrobutyrine, D-erythro-β-methylaspartic acid, and L-arginine. A number of structural variants of nodularin have been identified with varying toxicity [7]. Modifications to the ADDA moiety and the D-glutamic acid significantly alter the toxicity of the variants compared to nodularin.

The acid-base properties of nodularin have not been studied, but although the acid-base properties have not been identified, the molecule shares many similarities with microcystins. Like microcystin-LR, nodularin possesses two ionizable carboxyl groups and one ionizable amine group. It is reasonably expected, therefore, that nodularin will exist as a zwitterion in solution at environmentally relevant pHs, such as is the case for microcystin-LR. The pK_a value for microcystin has been estimated as 3.3. A similar value is likely for nodularin. At environmentally relevant pH values (between 4 and 10), therefore, nodularin is expected to have a net negative charge owing to the deprotonation of the two carboxyl groups and protonation of the amine group. The ADDA residue imparts hydrophobicity to the molecule.

3.2.2 Acute and Chronic Toxicity of Nodularin

Nodularin is considered a potent hepatotoxin but is generally thought to be less toxic than the similarly structured microcystins. Symptoms of exposure to nodularin include weakness, pallor, cold extremities, labored breathing, vomiting, diarrhea, and disruption of liver function. Much less is known about the health effects of nodularin compared to some of the other HAB toxins. However, some information has begun to emerge regarding the acute toxicity of this compound (for recent reviews of the toxicity of nodularin, see [7, 8]). Little is known, for example, regarding the various modes of uptake into the body, and little or no detailed information is available regarding the uptake of nodularin via dermal, oral, or inhalation pathways. A review of the literature [8, 9] showed that nodularin bioaccumulates in a variety of species, including fish, birds, shrimp, zooplankton, and benthic organisms. The toxin may bioaccumulate in fish, for example, through the ingestion of copepods or fecal pellets from copepods [8].

Like microcystin, nodularin is a potent hepatotoxin with an intraperitoneal LD_{50} (median lethal dose) value in mice of 50 mg/kg. No data are available regarding the oral LD_{50} for nodularin, but it is expected the value would be comparable to other HAB toxins (see Table 3.1). Pearson et al. [7] reviewed the toxicology research on nodularin and described how the toxin inhibits the function of protein phosphatase 1 (PP1) and protein phosphatase 2A (PP2A). This mechanism is similar to the toxic effect induced by microcystins. However, nodularin apparently does not bind covalently to PP enzymes, and microcystins do. The protein phosphatase enzymes play a critical role in a host of cellular processes, such as glycogen control, muscle contraction, and cell progression, to name just a few. As a result, nodularin disrupts a variety of cell functions. It is believed that both the ADDA moiety and the N-methyldehydrobutyrine group on nodularin are involved in disrupting the function of PP1 and PP2A. The ADDA moiety can bind to the hydrophobic groove of PP1, and the N-methyldehydrobutyrine group on nodularin binds to PP2A. The binding of nodularin to these proteins prevents the binding of substrate molecules.

The ecotoxicity of nodularin has been reviewed, with the indication that this compound accumulates and is toxic to a variety of aquatic species, including copepods, fish, shellfish, and others [10]. In fact, the decreased abundance of fish and crab in southwestern Australia is suspected to be caused by blooms of toxic *Nodularia* [11]. Anecdotal evidence from interviews with local fishers suggested that fish catches were reduced in areas near a *Nodularia* bloom. Fishers also noted visually observing greater numbers of dead fish near areas of the bloom. These observations were indirectly

TABLE 3.1

Basic Properties of the HAB Toxins

Toxin	MW	Formula	pK_a	Hydrophobicity
Microcystin-LR	995	$C_{49}H_{74}N_{10}O_{12}$	3.3	Hydrophobic
Nodularin	825	$C_{41}H_{60}N_8O_{10}$	~3.3	Hydrophobic
Cylindrospermopsin	415	$C_{15}H_{21}N_5O_7S$	8.8	Hydrophilic
Anatoxin-a	165	$C_{10}H_{15}NO$	9.6	—
Saxitoxin	299	$C_{10}H_{17}N_7O_4$	8.22, 11.28	—

supported by chlorophyll measurements as a proxy for *Nodularia*, as lower fish abundance was found in regions of the estuary that had higher chlorophyll levels. A number of studies have documented the uptake of nodularin in a variety of fish species, including herring, flounder, and trout [10].

Little or no information is available regarding the health impacts associated with short-term or chronic exposure to nodularin.

3.2.3 Genotoxicity and Carcinogenicity of Nodularin

Nodularin is known to be genotoxic and is a suspected carcinogen (for a review of the genotoxicity and carcinogenicity of nodularin, see [12]). Both in vitro and in vivo studies have shown nodularin to be genotoxic. Nodularin exerts a genotoxic effect via a number of mechanisms, namely, inducing DNA strand breakage, chromosomal damage, and oxidative stress. In vitro studies suggested nodularin does not form DNA adducts, but it can induce DNA strand cleavage. It has been demonstrated that nodularin is capable of inducing the production of reactive oxygen species (ROS), which subsequently results in oxidative stress. Nodularin is also known to be clastogenic and can disrupt or break sections of chromosome. In addition to genotoxicity, nodularin has been demonstrated to be a carcinogen through the inhibition of PP1 and PP2A. Apparently, the presence of nodularin can disrupt the regulation of hepatocytes in the cell.

3.2.4 Skin Irritation and Other Effects Induced by *Nodularia* and Nodularin

Epidemiological data demonstrate the effects of *Nodularia* on the skin and respiratory system. As described, in 1991 eight people became sick with a variety of symptoms following exposure to *Nodularia* in Murray River, Australia. The main symptoms included skin and eye irritation and respiratory issues. The respiratory issues included sore throat, hay fever, and asthma.

3.3 Microcystins

There have been a number of accounts identifying microcystins in the poisoning of humans and animals over the last 50 years. In one of the earliest studies, *Microcystis* was linked to an outbreak of gastrointestinal problems in Saskatchewan, Canada, in the early 1960s [2]. A number of people were exposed to *Microcystis* on recreational contact with a contaminated water body. Symptoms experienced included nausea, diarrhea, and headaches. No deaths occurred as a result of the exposure. Stool samples from exposed individuals tested positive for the presence of *Microcystis*.

In the 1990s, over 1000 people are believed to have become ill as result of exposure to *Microcystis*. There were 1118 cases of diarrhea recorded in a population of individuals following the flooding of the Itiparica Dam in Bahio, Brazil [2]. Most (>70%) of the recorded cases of diarrhea were in children 5 years of age or less. Although *Microcystis* and *Anabaena* were detected in the untreated water, the organisms were not found above detection levels in the treated drinking water. The coincident outbreak and reservoir flooding, along with the identification of causative organisms in the untreated water, suggested some possible role of these organisms in the observed symptoms.

An outbreak at a dialysis clinic in Caruaru, Brazil, is believed to be caused by the presence of *Microcystis*. Those affected were 116 of 131 dialysis patients, and 100 experienced acute liver failure. Both microcystin and cylindrospermopsin were identified in the contaminated water. Only microcystin, however, was identified in the stools of the dialysis patients affected, although the analytical methodology used to isolate the toxins from stool samples was not definitive. Therefore, the role cylindrospermopsin may have played in this outbreak is not entirely clear.

3.3.1 Microcystin Properties

Over 80 congeners have been identified for microcystin. This chapter focuses on the properties of microcystin-LR with important distinctions made for other congeners when appropriate. The chemical formula for microcystin-LR is $C_{49}H_{74}N_{10}O_{12}$, and the toxin has a molecular weight of 995.17. Microcystin-LR is one of the larger HAB toxins and is slightly larger than nodularin. The chemical structure of microcystin-LR is shown in Figure 3.3. Microcystins, as a class of compounds, are cyclic heptapeptides containing the β-amino acid ADDA, D-alanine, D-methyl aspartate, D-glutamic acid, N-methyldehydroalanine (Mdha), and two variable amino acids. In

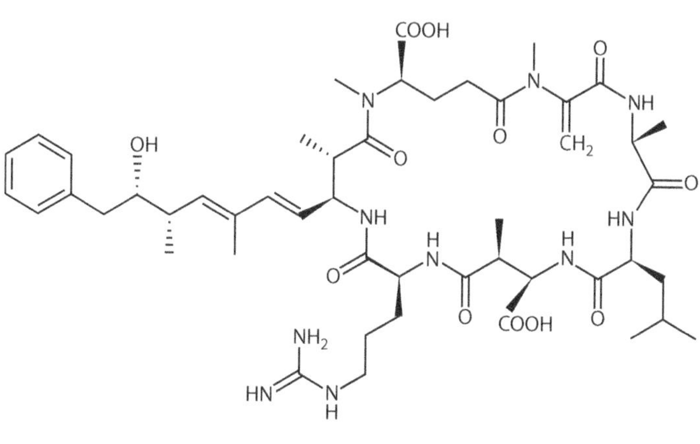

FIGURE 3.3
Chemical structure of microcystin-LR.

microcystin-LR, the two variable amino acids are L-arginine and L-leucine. The microcystins are similar in structure to nodularin, except that the microcystins contain seven amino acid groups, and nodularin contains five. The different amino acid side chains can alter physical and chemical properties, as well as toxicity, of the various microcystin congeners.

The acid-base properties of microcystin-LR have been more thoroughly studied [13] compared to other HAB toxins. Microcystin-LR possesses two ionizable carboxyl groups and one ionizable amine group. The pK_a values for the carboxyl groups are 2.09 and 2.19. The pK_a for the amine group is 12.48 [14]. At very low pH values (<2), therefore, the microcystin-LR molecule has a net positive charge. At these very low pH conditions, the amine group contributes one positive charge to the molecule, and the two carboxyl groups are uncharged because of their protonation. At pH values between the pK_a values for the carboxyl groups, microcystin-LR is a zwitterion with no net charge. Between pH values of 2.09 and 2.19, the amine group will contribute one positive charge to the molecule, and one carboxyl group will contribute a negative charge. The second carboxyl group will remain uncharged because of the higher pK_a value. At pH values higher than 2.19, microcystin-LR will have a net negative charge because of the one positive charge contributed by the amine group and the two negative charges provided by the carboxyl groups. The amine group will lose a proton at only very high pH values (>12.48). Under most environmentally relevant pH conditions, therefore, microcystin-LR behaves as a zwitterion with a net negative charge of –1. The overall pK_a value for microcystin-LR has been estimated as 3.3 [15]. In addition to controlling the net charge of microcystin-LR, the acid-base properties provide for the binding of microcystin-LR with Fe(III) [13].

Microcystins are relatively hydrophobic compounds, and the hydrophobicity depends on pH [14]. The hydrophobicity of microcystin-LR, for example, has been shown to decrease with increasing pH, from a log DOw value of more than 2 at pH values less than 2 to a value of close to –2 at a pH of 10. The lower hydrophobicity of microcystin-LR at higher pH values suggests that the tendency of the compound to partition into aquatic organisms will decrease at the higher pH values.

3.3.2 Acute and Chronic Toxicity of Microcystins

Like nodularin, microcystins are potent hepatotoxins. In fact, microcystins are generally considered more toxic than nodularin. In 2006, the USEPA did a comprehensive review of the literature to determine the health effects of microcystin-LR, -RR, -YR, and -LA as part of the contaminant review process under the Safe Drinking Water Act [2]. Primary symptoms associated with acute exposure to microcystin-LR include nausea, stomach pain, diarrhea, headache, and eye, ear, and throat irritation.

TABLE 3.2

Common Classes of HAB Toxins and Their Acute Toxicity

Toxin	Target	Oral LD_{50} (mg/kg)	NOAEL (mg/kg/day)	Subchronic RfD (mg/kg/day)
Nodularin	Liver toxin	—	—	—
Microcystin-LR	Liver toxin	5	0.040	6×10^{-6}
Cylindrospermopsin	Liver, kidney	4.4–6.9	0.03	3×10^{-5}
Anatoxin-a	Neurotoxin	16.2	0.5	5×10^{-4}
Saxitoxin	Neurotoxin	0.263	0.5 µg/kg/day	

The oral LD_{50} value for microcystin-LR in mice is 5 mg/kg, which is one of the lower LD_{50} values for the various HAB toxins (see Table 3.2). Like nodularin, microcystin-LR inhibits the function of PP1 and PP2A. Unlike nodularin, however, microcystin-LR apparently can bind covalently to protein phosphatase enzymes. As discussed, the protein phosphatase enzymes play a critical role in a host of cellular processes, such as glycogen control, muscle contraction, and cell progression, to name just a few. As a result, nodularin disrupts a variety of cell functions. It is believed that both the ADDA moiety and the N-methyldehydrobutyrine group on microcystin-LR are involved in disrupting the function of PP1 and PP2A. Similarly, as discussed for nodularin, the ADDA moiety can bind to the hydrophobic groove of PP1, and the N-methyldehydrobutyrine group on nodularin binds to PP2A. The binding of nodularin to these proteins prevents the binding of substrate molecules. Disruption of PP1 and PP2A leads to retraction of the hepatocyte cells, which results in pooling of blood in the liver and organ failure.

A number of studies have documented the impacts of *Microcystis* and microcystins on aquatic plants and animals and other organisms (for a review, see [10]). *Microcystis* has effects on both lower-order and higher-order organisms. *Microcystis* can affect, for example, the growth of other algae species. Studies have shown that the presence of *Microcystis* results in a reduction in growth of other algae because of photosynthesis inhibition. *Microcystis* has also been shown to retard the growth of a number of aquatic plants, including duckweed, *Ceratophyllum demersum*, and *Spirodela ologorrhiza*. *Microcystis* apparently reduces the amount of chlorophyll a and b and may also reduce growth through oxidative stress. *Microcystis* has also been shown to have a negative impact on the growth of zooplankton. Although bivalves, crabs, and crayfish can potentially bioaccumulate microcystins, these species seem to be relatively insensitive to the toxic effects of the compound. However, a number of massive fish kills have been seen as a result of *Microcystis*, which appears to affect the liver and kidneys of fish, similar to the toxic effects in mammals. Studies have shown toxic effects of *Microcystis* on carp, tilapia, trout, and other species.

3.3.3 Genotoxicity and Carcinogenicity of Microcystins

Microcystin-LR is genotoxic and is a suspected human carcinogen (for a review of the genotoxicity and carcinogenicity of microcystin-LR, see [12]). Both in vitro and in vivo studies have shown microcystin-LR to be genotoxic. Like nodularin, microcystin-LR exerts a genotoxic effect via a number of mechanisms, namely, inducing DNA strand breakage, chromosomal damage, and oxidative stress.

3.3.4 Skin Irritation and Other Effects Induced by *Microcystis* and Microcystins

A number of cases of *Microcystis* exposure during training exercises have been documented in the literature. Codd et al. [16], for example, documented incidences of rashes, blistering mouths, asthma, coughing, and vomiting in soldiers performing canoe-rolling drills during training in different parts of England.

3.4 Anatoxin-a

The first suspected case of human death attributable to anatoxin-a occurred in Dane County, Wisconsin, in 2002. A summary of the event is given in the USEPA toxicological review of anatoxin-a [1]. Near dusk, on a 97°F July day, a 17-year-old boy went swimming with four friends in a scum-covered pond adjacent to a golf course. The boys were "roughhousing" and as a result ingested significant volumes of pond water. Approximately 48 hours later, the boy experienced seizures, went into shock, and eventually died as a result of heart failure. Another of the teenage boys experienced severe diarrhea and stomach pain. The three others present that day experienced only minor symptoms. Blood samples collected from the boys were positive for *Anabaena flos-aquae* and anatoxin-a, suggesting this toxic substance was implicated in the death of the boy and illness experienced by others. Since the initial assessment, it has been reported that anatoxin-a may not have been the main cause of the fatality because of the prolonged period of time between exposure and the presentation of acute symptoms. Follow-up studies also determined that the compound originally thought to be anatoxin-a in body fluids and liver tissue samples was in fact the amino acid phenylalanine.

Concern about anatoxin-a occurred decades earlier, in 1961, however, when a number of cows died in Saskatchewan, Canada, following ingestion of pond water contaminated with *Anabaena flos-aquae* (for a summary of the account, see [1]). The connection between the deaths and *Anabaena flos-aquae* was established by identifying *Anabaena flos-aquae* in pond scum samples. Researchers also isolated the toxin from the sample and found it resulted

FIGURE 3.4
The structure of anatoxin-a.

in similar symptoms as those observed and subsequently termed the toxin "very fast death factor." In 1966, the structure of this toxin was identified, and the compound was later renamed anatoxin-a.

3.4.1 Anatoxin-a Properties

The chemical formula for anatoxin-a is $C_{10}H_{15}NO$. The toxin has a molecular weight of 165.23. As mentioned, the chemical structure of anatoxin-a was first elucidated in 1966 [1]. The primary chemical structure of anatoxin-a is shown in Figure 3.4. The toxin has been characterized as having a semirigid bicycle secondary amine structure. Anatoxin-a exists as two enantiomers, but only (+) anatoxin-a is found in nature. Three other naturally occurring structural variants of anatoxin-a have been identified: dihydroanatoxin-a, homoanatoxin-a, and dihydrohomoanatoxin-a.

The electrostatic character of anatoxin-a stems from the protonation of the nitrogen group. The pK_a value for anatoxin-a has been estimated as 9.6 [17]. At pH values less than 9.6, the nitrogen group is in the protonated state and has a positive charge. Therefore, anatoxin-a exists as a positively charged ion at environmentally relevant pH values. At pH values higher than 9.6, the compound is deprotonated and has no net charge.

3.4.2 Acute and Chronic Toxicity of Anatoxin-a

Much less is known about the effect of anatoxin-a on human health compared to the available data on microcystins. Some information, however, is available regarding the acute toxicity of this compound and other health effects (for recent reviews of the toxicity of anatoxin-a, see [1]). Although an understanding of the acute toxicity of anatoxin-a is beginning to emerge, little is known regarding the various modes of exposure. Little or no detailed information is available, for example, regarding the uptake of anatoxin-a via dermal, oral, or inhalation pathways. Although data are scarce, a few studies have demonstrated that anatoxin-a bioaccumulates in fish [8]. Little or no information is available regarding the metabolism and elimination of anatoxin-a in humans or animals.

Anatoxin-a is a known neurotoxin and affects both the peripheral and the central nervous system. Exposure to anatoxin-a results in muscular twitching, loss of coordination, respiratory paralysis, and death. Anatoxin-a is acutely toxic, with an oral LD_{50} in mice of 16.2 mg/kg (see Table 3.2). Anatoxin-a is nicotinic acetylcholine receptor agonist because it mimics the action of acetylcholine at neuromuscular nicotine receptors [1]. Anatoxin-a has similar stereospecificity as acetylcholine and nicotine. Anatoxin-a possesses the essential structural form of an acetylcholine receptor agonist, that is, a planar hydrogen bond located 5.9 Å from a bulky cationic group. Unlike acetylcholine, however, anatoxin-a is resistant to enzymatic degradation. Although acetylcholine is degraded by acetylcholinesterase, which prevents overstimulation, anatoxin-a has no such regulation. The presence of anatoxin-a therefore results in overstimulation at the nicotine receptors. Anatoxin-a is, in fact, a 3.6 times more potent stimulator than acetylcholine and 7–136 times more potent than nicotine.

Only a few studies have documented the acute toxicity of anatoxin-a to aquatic organisms [8]. The acute toxicity of anatoxin-a to the copepods *Diaptomus birgei* and *Daphnia pulicaria*, for example, was demonstrated by Demott et al. [18]. Anatoxin-a was also shown to be acutely toxic to the rotifer *Brachionus calyciflorus* [8].

Comparatively much less is known about the health effects to humans and animals associated with short-term and chronic exposure to anatoxin-a. Based on 28-day studies with mice, the short-term no observed adverse effect level (NOAEL) for anatoxin-a was established by USEPA as 2.5 mg/kg/day [1]. A subchronic NOAEL was established as 0.5 mg/kg/day. Using an uncertainty factor of 1000 to account for interspecies variability and other factors, the USEPA established a subchronic reference dose (RfD) value for anatoxin-a of 5×10^{-4} mg/kg/day. No information is currently available regarding the health effects caused by chronic exposure to anatoxin-a.

3.4.3 Genotoxicity and Carcinogenicity of Anatoxin-a

Little or no information is available regarding the genotoxicity and cancer-causing potential of anatoxin-a [1]. It is generally believed the acute neurotoxic effects of anatoxin-a are of greatest concern for human health.

3.4.4 Skin Irritation Induced by Anatoxin-a

No information is currently available regarding the health effects associated with dermal exposure to anatoxin-a.

3.5 Cylindrospermopsin

The first documented outbreak of *Cylindrospermopsis* occurred on Palm Island, Queensland, Australia, in 1979 [19]. The outbreak (or so-called mystery disease) occurred following the application of copper sulfate to control an algal bloom in Solomon Dam and resulted in the hospitalization of over 138 aboriginal children (median age of 8.4 years) and 10 adults. It was suspected that the addition of copper sulfate disrupted *Cylindrospermopsis* cells, which facilitated the release of the toxin cylindrospermopsin. Symptoms experienced by the children and adults included fever, headache, vomiting, bloody diarrhea, hepatomegaly, and renal damage. It should be noted that an alternative explanation for the outbreak has also been posited. It has been suggested, for example, that the outbreak may have instead been the result of copper sulfate poisoning [20]. Apparently, exposure to high levels of copper sulfate may produce some similar symptoms as those observed. It has come to light that the contractor hired to control the bloom was not qualified and possibly overdosed the reservoir. Whether or not cylindrospermopsin was the causative agent in the outbreak on Palm Island, significant research since has established the serious health risks posed by this toxin.

Interestingly, the outbreak at Palm Island led to speculation that cylindrospermopsin might be responsible for other, similar outbreaks in Australia. Accounts of Barcoo fever, for example, date to the late 1800s. Barcoo fever is an illness known to occur in the Australian outback and presents with symptoms similar to those observed in Palm Island, including vomiting, fever, and myalgia. A large outbreak of Barcoo fever is suspected to have occurred in the town of Toowoomba, in South-East Queensland, Australia, in 1903. The aboriginal population avoided the disease by not directly drinking surface water, but instead taking water for drinking from pits dug in the sand of the stream bed. Interestingly, recent research suggests little retention of cylindrospermopsin in sandy sediment [21]; however, the presence of organic carbon in sediment may provide for microbial degradation.

An outbreak at a dialysis clinic in Caruaru, Brazil, may also have been related to the consumption of cylindrospermopsin. As discussed, 116 of 131 dialysis patients were affected, and 100 experienced acute liver failure. Both microcystin and cylindrospermopsin were identified in the contaminated water. Only microcystin, however, was identified in the stools of the dialysis patients affected, although the analytical methodology used to isolate the toxins from stool samples was not definitive. Therefore, the role cylindrospermopsin may have played in this outbreak is not entirely clear.

3.5.1 Cylindrospermopsin Properties

The chemical formula for cylindrospermopsin is $C_{15}H_{21}N_5O_7S$, and the toxin has a molecular weight of 415.43. The chemical structure of cylindrospermopsin was first elucidated in 1992 by Ohtani et al. [22] using a combination of nuclear magnetic resonance (NMR) and mass spectrometry (MS) and is shown in Figure 3.5. The toxin consists of two main parts: a tricyclic guanidine moiety and a hydroxyl methyl uracil group. Two naturally occurring

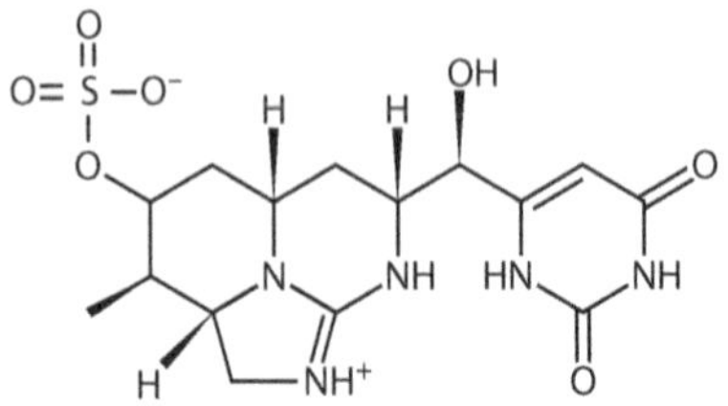

FIGURE 3.5
The structure of cylindrospermopsin.

structural variants of cylindrospermopsin have been identified: 7-epiCYN and 7-deoxy-CYN.

Cylindrospermopsin is zwitterionic, given the negatively charged sulfate group attached to C-12 and the positive charge on the nitrogen of the guanidine moiety. The pK_a value for cylindrospermopsin has been estimated as 8.8 [23]. Therefore, at a pH value of close to 8.8, cylindrospermopsin has a net neutral charge, with equal amounts of positive and negative charge. At pH values less than 8.8, cylindrospermopsin has a net positive charge, with the positively charged nitrogen on the guanidine moiety contributing more to the total net charge than the negatively charged sulfate group. At pH values greater than 8.8, cylindrospermopsin has a net negative charge, with the charge behavior dominated by the sulfate group.

3.5.2 Acute and Chronic Toxicity of Cylindrospermopsin

A comprehensive understanding of the effect of cylindrospermopsin on human health has not yet been developed, but some information has begun to emerge regarding the acute toxicity of this compound (for recent reviews of the toxicity of cylindrospermopsin, see [24–26]). Little is known, for example, regarding the various modes of exposure. Little or no detailed information is available regarding the uptake of cylindrospermopsin via dermal, oral, or inhalation pathways. Interestingly, a number of studies have shown that cylindrospermopsin may bioaccumulate in aquatic plants and animals, despite the hydrophilic nature of the compounds [27]. In general, bioaccumulation has been found to decrease with increased complexity of the organism. The identification of cylindrospermopsin in urine suggests the compound is not effectively metabolized in mammals.

Cylindrospermopsin is a known hepatotoxin and may also be toxic to other organs, including the thymus, kidneys, spleen, and heart. Most of the in vivo toxicological studies have been carried out with mice. Cylindrospermopsin is acutely toxic, with an oral LD_{50} in mice between 4.4 and 6.9 mg/kg, which is comparable to other HAB toxins, such as certain congeners of microcystin (see Table 3.1). The uracil moiety has been shown to play an important role in establishing the toxicity of cylindrospermopsin because cleavage of this group using chlorine greatly reduced toxicity [28]. Although a number of mechanisms may be responsible for the acute toxicity of cylindrospermopsin in mammals, both in vitro and in vivo studies indicated the primary mode of action is through the inhibition of protein synthesis. Cylindrospermopsin appears to inhibit the production of glutathione, with increasing inhibition observed with increasing exposure to the toxin. The formation of p450 metabolites may also play a role in the toxic action of cylindrospermopsin.

Studies have also documented the acute toxicity of cylindrospermopsin to a variety of aquatic species, including fish, amphibians, bivalves, and plants. The toxicity of cylindrospermopsin to the tadpoles of the cane toad, *Bufo marinus*, for example, was demonstrated by White et al. [29]. They found that 66% of the tadpoles died on exposure to *Cylindrospermopsis* cell extracts and cylindrospermopsin at 232 µg/L. Puerto et al. demonstrated the toxic effects of cylindrospermopsin on aquatic bivalves, including the Mediterranean mussels and the Asian clam [30]. Berry et al. showed that cylindrospermopsin is lethal to zebrafish embryos [31]. Cylindrospermopsin has also been shown to disrupt the growth of aquatic plants, including duckweed, water thyme, white mustard, as well as tabaco. It is believed that cylindrospermopsin disrupts proteolysis in plants. Cylindrospermopsin has also been shown to be toxic to planktonic organisms, such as *Daphnia magma*.

Comparatively much less is known about the health effects to humans and animals associated with short-term and chronic exposure to cylindrospermopsin.

3.5.3 Genotoxicity and Carcinogenicity of Cylindrospermopsin

Cylindrospermopsin is suspected to be genotoxic and possibly carcinogenic; however, further evidence is needed to more fully substantiate these effects as well as elucidate mode of action (for a recent review, see [12]). A number of in vitro and in vivo studies have shown cylindrospermopsin to be genotoxic. Cylindrospermopsin may form DNA adducts and is involved in DNA strand cleavage. There is comparatively less data available regarding whether cylindrospermopsin acts as a carcinogen. Some evidence suggests, however, that cylindrospermopsin may promote tumor formation in mice. These last studies, however, were based on a limited set of data and therefore were found not statistically robust. Further research is needed to better substantiate the genotoxicity and carcinogenicity of cylindrospermopsin.

3.5.4 Skin Irritation Induced by *Cylindrospermopsis* and Cylindrospermopsin

A study was carried out on healthy human volunteers to identify the potential for skin irritation on exposure to *Cylindrospermopsis* cells [32]. Human volunteers were exposed to both whole and lysed cells of cylindrospermopsin, via a small skin patch, at concentrations of 5000 to 200,000 cells/mL. The results of the study showed that only 6–18% of volunteers showed a mild reaction; the remaining volunteers were unaffected. For those volunteers who experienced a mild reaction, the condition cleared up without treatment.

3.6 Saxitoxins

There are no reported cases of exposure to STX in freshwater systems. However, given the increasing development of seawater and brackish water desalination, potential exposure from drinking water will increase. STX has historically been of greatest concern with respect to the consumption of shellfish as it causes paralytic shellfish poisoning (PSP). The prevalence of HABs in marine systems and the potential for PSP is widespread. Shellfish fisheries are closed on a routine basis because of PSP.

3.6.1 Saxitoxin Properties

The chemical formula for STX is $C_{10}H_{17}N_7O_4$. The toxin has a molecular weight of 299.29 (see Figure 3.6). Three other naturally occurring structural variants of STX have been identified: neosaxitoxin (NSTX), gonyautoxin (GTX), and decarbamoylsaxitoxin (dcSTX). STX is believed to have a positive charge under most environmentally relevant pH conditions. Two pK_a values have been identified for STX, at 8.22 and 11.28 [33, 34]. At a pH value less than 8.22, STX has two positive charges per molecule. Between pH values of 8.22 and 11.28, the molecule is expected to have one positive charge and be uncharged at pH values above 11.28.

3.6.2 Acute and Chronic Toxicity of Saxitoxin

Saxitoxin is a potent neurotoxin and causes the condition PSP [7]. The primary symptoms of exposure to STX include burning of the lips, throat, and tongue and numbness of the face. Persons exposed may also experience

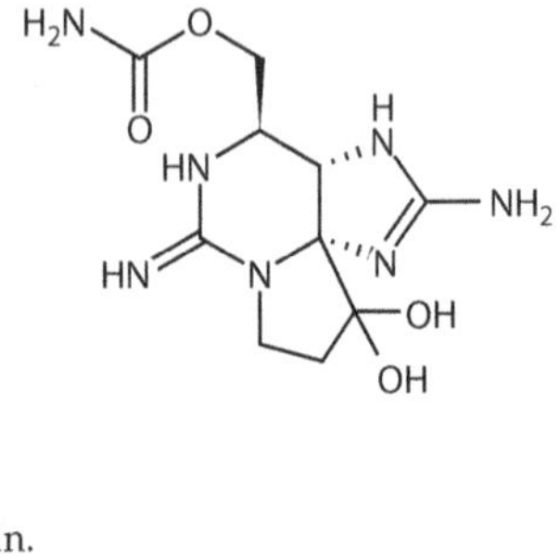

FIGURE 3.6
Chemical structure of saxitoxin.

perspiration, vomiting, and diarrhea. In more severe cases of exposure to STX, the numbness spreads to the neck, and the exposed individual experiences muscle weakness, loss of motor coordination, and ultimately paralysis. Paralysis leads to cardiovascular failure and death. It is believed that STX acts by blocking voltage-gated sodium and calcium channels in neural cell membranes. The guanidinium group and C12 hydroxyl play a key role in the binding of STX to the channel proteins.

Like the other HAB toxins, STX is acutely toxic, with an intraperitoneal LD_{50} in mice of 0.10 mg/kg and an oral LD_{50} of 0.263 mg/kg. The LD_{50} in humans has been determined to be 0.0057 mg/kg. Comparing the LD_{50} values of the various HAB toxins shows that STX is the most acutely toxic. Structural variants of STX act in a similar fashion to STX but are generally less toxic.

Given the concern over PSP and consumption of seafood, a significant number of studies have examined the effect of STX on aquatic organisms [10]. STX is known to bioaccumulate in copepods, clams, crabs, mussels, and other marine organisms. Marine organisms appear to be relatively insensitive to the toxic effects of STX but serve to transmit the toxins up the food chain.

3.6.3 Genotoxicity and Carcinogenicity of Saxitoxin

Currently, there are no data available regarding the subchronic, genotoxic, or carcinogenic potential on exposure to STX.

3.6.4 Skin Irritation Induced by Saxitoxin

No studies have documented skin irritation or other effects caused by dermal exposure to STX in freshwater systems.

References

1. US Environmental Protection Agency, *Toxicological Reviews of Cyanobacterial Toxins: Anatoxin-A*. National Center for Environmental Assessment, Cincinnati, OH, 2006, p. 34.
2. US Environmental Protection Agency, *Toxicological Reviews of Cyanobacterial Toxins: Microcystins LR, RR, YR and LA*. National Center for Environmental Assessment, Cincinnati, OH, 2006, p. 226.
3. G. Francis, Poisonous Australian lake. *Nature* (London), 18 (1878) 11–12.
4. T. Heresztyn and B.C. Nicholson, Nodularin concentrations in Lakes Alexandrina and Albert, South Australia, during a bloom of the cyanobacterium (blue-green alga) *Nodularia spumigena* and degradation of the toxin. *Environmental Toxicology and Water Quality*, 12 (1997) 273–282.
5. P.R. Hunter, Cyanobacterial toxins and human health. *Journal of Applied Microbiology*, 84 (1998) 35S-40S.

6. F.S. Soong, E. Maynard, and K. Kirke, Illness associated with blue-green algae. *Medical Journal of Australia*, 156 (1992) 67.
7. L. Pearson, T. Mihali, M. Moffitt, R. Kellmann, and B. Neilan, On the chemistry, toxicology and genetics of the cyanobacterial toxins, microcystin, nodularin, saxitoxin and cylindrospermopsin. *Marine Drugs*, 8 (2010) 1650–1680.
8. A.d.S. Ferrão-Filho and B. Kozlowsky-Suzuki, Cyanotoxins: bioaccumulation and effects on aquatic animals. *Marine Drugs*, 9 (2011) 2729–2772.
9. M. Karjalainen, J.-P. Pääkkönen, H. Peltonen, V. Sipiä, T. Valtonen, and M. Viitasalo, Nodularin concentrations in Baltic Sea zooplankton and fish during a cyanobacterial bloom. *Marine Biology*, 155 (2008) 483–491.
10. C. Wiegand and S. Pflugmacher, Ecotoxicological effects of selected cyanobacterial secondary metabolites a short review. *Toxicology and Applied Pharmacology*, 203 (2005) 201–218.
11. I.C. Potter, N.R. Loneragan, R.C.J. Lenanton, and P.J. Chrystal, Blue-green algae and fish population changes in a eutrophic estuary. *Marine Pollution Bulletin*, 14 (1983) 228–233.
12. B. Žegura, A. Štraser, and M. Filipič, Genotoxicity and potential carcinogenicity of cyanobacterial toxins—a review. *Mutation Research/Reviews in Mutation Research*, 727 (2011) 16–41.
13. A.R. Klein, D.S. Baldwin, and E. Silvester, Proton and iron binding by the cyanobacterial toxin microcystin-LR. *Environmental Science and Technology*, 47 (2013) 5178–5184.
14. P.G.-J. de Maagd, A.J. Hendriks, W. Seinen, and D.T.H.M. Sijm, pH-dependent hydrophobicity of the cyanobacteria toxin microcystin-LR. *Water Research*, 33 (1999) 677–680.
15. C. Rivasseau, S. Martins, and M.-C. Hennion, Determination of some physico-chemical parameters of microcystins (cyanobacterial toxins) and trace level analysis in environmental samples using liquid chromatography. *Journal of Chromatography A*, 799 (1998) 155–169.
16. G. Codd, S. Bell, K. Kaya, C. Ward, K. Beattie, and J. Metcalf, Cyanobacterial toxins, exposure routes and human health. *European Journal of Phycology*, 34 (1999) 405–415.
17. S. Klitzke, C. Beusch, and J. Fastner, Sorption of the cyanobacterial toxins cylindrospermopsin and anatoxin-a to sediments. *Water Research*, 45 (2011) 1338–1346.
18. W.R. DeMott, Q.-X. Zhang, and W.W. Carmichael, Effects of toxic cyanobacteria and purified toxins on the survival and feeding of a copepod and three species of *Daphnia*. *Limnology and Oceanography*, 36 (1991) 1346–1357.
19. D.J. Griffiths and M.L. Saker, The Palm Island mystery disease 20 years on: a review of research on the cyanotoxin cylindrospermopsin. *Environmental Toxicology*, 18 (2003) 78–93.
20. P. Prociv, Algal toxins or copper poisoning—revisiting the Palm Island "epidemic." *Medical Journal of Australia*, 181 (2004) 344.
21. S. Klitzke, S. Apelt, C. Weiler, J. Fastner, and I. Chorus, Retention and degradation of the cyanobacterial toxin cylindrospermopsin in sediments—the role of sediment preconditioning and DOM composition. *Toxicon*, 55 (2010) 999–1007.
22. I. Ohtani, R.E. Moore, and M.T.C. Runnegar, Cylindrospermopsin: a potent hepatotoxin from the blue-green alga *Cylindrospermopsis raciborskii*. *Journal of the American Chemical Society*, 114 (1992) 7941–7942.

23. G.D. Onstad, S. Strauch, J. Meriluoto, G.A. Codd, and U. von Gunten, Selective oxidation of key functional groups in cyanotoxins during drinking water ozonation. *Environmental Science and Technology*, 41 (2007) 4397–4404.

24. D.M. Evans and P.J. Murphy, The cylindrospermopsin alkaloids, in: K. Hans-Joachim (Ed.), *The Alkaloids: Chemistry and Biology*. New York: Academic Press, 2011, pp. 1–77.

25. C. Moreira, J. Azevedo, A. Antunes, and V. Vasconcelos, Cylindrospermopsin: occurrence, methods of detection and toxicology. *Journal of Applied Microbiology*, 114 (2013) 605–620.

26. B. Poniedziałek, P. Rzymski, and M. Kokociński, Cylindrospermopsin: water-linked potential threat to human health in Europe. *Environmental Toxicology and Pharmacology*, 34 (2012) 651–660.

27. S. Kinnear, Cylindrospermopsin: a decade of progress on bioaccumulation research. *Marine Drugs*, 8 (2010) 542–564.

28. R. Banker, S. Carmeli, M. Werman, B. Teltsch, R. Porat, and A. Sukenik, Uracil moiety is required for toxicity of the cyanobacterial hepatotoxin cylindrospermopsin. *Journal of Toxicology and Environmental Health, Part A*, 62 (2001) 281–288.

29. S.H. White, L.J. Duivenvoorden, L.D. Fabbro, and G.K. Eaglesham, Mortality and toxin bioaccumulation in *Bufo marinus* following exposure to *Cylindrospermopsis raciborskii* cell extracts and live cultures. *Environmental Pollution*, 147 (2007) 158–167.

30. M. Puerto, A. Campos, A. Prieto, A. Cameán, A.M.d. Almeida, A.V. Coelho, and V. Vasconcelos, Differential protein expression in two bivalve species, *Mytilus galloprovincialis* and *Corbicula fluminea*, exposed to *Cylindrospermopsis raciborskii* cells. *Aquatic Toxicology*, 101 (2011) 109–116.

31. J.P. Berry, P.D.L. Gibbs, M.C. Schmale, and M.L. Saker, Toxicity of cylindrospermopsin, and other apparent metabolites from *Cylindrospermopsis raciborskii* and *Aphanizomenon ovalisporum*, to the zebrafish (*Danio rerio*) embryo. *Toxicon*, 53 (2009) 289–299.

32. L. Pilotto, P. Hobson, M.D. Burch, G. Ranmuthugala, R. Attewell, and W. Weightman, Acute skin irritant effects of cyanobacteria (blue-green algae) in healthy volunteers. *Australian and New Zealand Journal of Public Health*, 28 (2004) 220–224.

33. R.S. Rogers and H. Rapoport, The pKa's of saxitoxin. *Journal of the American Chemical Society*, 102 (1980) 7335–7339.

34. Y. Shimizu, Chemistry and biochemistry of saxitoxin analogues and tetrodotoxina. *Annals of the New York Academy of Sciences*, 479 (1986) 24–31.

4

Regulation of HABs and HAB Toxins in Surface and Drinking Water

4.1 Introduction

To make significant progress in minimizing the occurrence of harmful algal blooms (HABs) and the impacts of HABs on drinking water, regulatory approaches are needed. Currently, there are no federal regulations or US guidelines for cyanobacteria or their toxins in both recreational and drinking water. However, a number of existing water quality regulations, at both the state and federal level, do provide means to manage and/or control HABs and their related toxins. In this chapter, regulations for HABs and HAB toxins in drinking water at the state, federal, and international levels are discussed. Other relevant regulations, such as the Harmful Algal Bloom and Hypoxia Research and Control Act, regulations managing HABs in recreational waters, as well as current attempts at both the state and federal levels to establish nutrient criteria, are examined.

4.2 Funding for Research and Risk Assessment on HABs and HAB Toxins

In 1993, a consortium of experts in marine sciences, fisheries, ecology, and other fields met at Woods Hole Oceanographic Institute (WHOI) to discuss emerging issues surrounding HABs in the United States. The meeting resulted in the development of a "national plan" to address the issue of marine biotoxins and harmful algae [1]. The national plan identified a number of major impediments for developing science-based solutions to the problems of marine biotoxins and HABs. The report also outlined a number of goals and objectives to begin to address these challenges. The major impediments to progress noted in the national plan included the lack of available information about biotoxins, the lack of reliable assays, a lack of understanding of the ecology and dynamics of HABs, a lack of information about the magnitude and extent of affected fisheries, and inadequate methods and warning systems to protect public health.

Although this national plan focused on marine toxins, the impediments were also relevant to freshwater systems and perhaps more so given the lack of research conducted at the time on HABs in freshwater lakes and reservoirs. Coastal waters are also increasingly being used as a source of drinking water as desalination technology becomes more economically feasible. Therefore, the issues and objectives identified in the national plan are of direct relevance to the increasing development of seawater desalination for enhancing drinking water resources.

Following the development of the national plan, a number of efforts in the United States have focused on establishing legislation to support research on HABs and HAB toxins, including freshwater HABs. The first such legislation was the Harmful Algal Bloom and Hypoxia Research and Control Act (HABHRCA) of 1998. The HABHRCA of 1998 authorized funding for the National Oceanic and Atmospheric Administration (NOAA) to establish a program focused on the research and control of HABs and hypoxia within oceans and estuaries of the United States, as well as the Great Lakes. According to NOAA, the focus of the program is "to advance the scientific understanding and ability to detect, monitor, assess, and predict HABs and to develop programs for research into methods of prevention, control, and mitigation of HABs [2]." The HABHRCA was reauthorized in 2004 and further mandated the development of five interagency reports on various aspects of HABs and HAB toxins. As a result of this legislation, the Federal Interagency Task Force on Harmful Algal Blooms, Hypoxia, and Human Health was established, and the task force published a number of scientific reports, including the following:

- *Harmful Algal Bloom Management and Response: Assessment and Plan* [3],
- *Scientific Assessment of Marine Harmful Algal Blooms* [4],
- *Scientific Assessment of Freshwater Harmful Algal Blooms* [5], and
- *Scientific Assessment of Hypoxia in U.S. Coastal Waters* [6].

The HABHRCA also authorized a number of major research studies, including the Ecology and Oceanography of Harmful Algal Blooms (ECOHAB) research program; the Prevention, Control, and Mitigation for Harmful Algal Blooms (PCM HAB) research program; the Monitoring and Event Response for Harmful Algal Blooms (MERHAB) research program; the Northern Gulf of Mexico Ecosystems and Hypoxia Assessment (NGOMEX) research program; and the Coastal Hypoxia Research Program (CHRP). A 2-year extension of the HABHRCA was passed by Congress in 2008, after which time several efforts were made to reauthorize the legislation. As of 2013, however, the program (and subsequently, NOAA) is not authorized to continue the programs established through the HABHRCA. The most recent version of the act (HABHRCA of 2014) was enacted on June 30, 2014 and authorizes up to $82 million over the period 2014 to 2018.

In 2009, efforts were made to pass the Freshwater Harmful Algal Bloom Research and Control Act (FHABRCA). Although the HABHRCA addressed HABs in the oceans and estuaries of the United States and the Great Lakes, it did not address other water bodies, including the numerous lakes and ponds in the United States that are affected by HABs at an increasing rate. The FHABRCA would authorize the US Environmental Protection Agency (USEPA) to develop a national HAB research plan and expand existing research efforts, such as MERHAB and ECOHAB, to consider other freshwater bodies. The FHABRCA has not been passed by Congress and signed into law. The most recent attempts at passage of the FHABRCA have incorporated the act into the HABHRCA reauthorization.

4.3 Drinking Water Regulations

In 2003, the World Health Organization (WHO) established a provisional drinking water guideline of 1 µg/L for microcystin-LR, reported as the sum of both cell-bound and free microcystin [7]. Microcystin congeners, other than -LR, were not included in the WHO guideline because of lack of sufficient information at the time of promulgation. The WHO provisional guideline for microcystin-LR was established primarily based on laboratory studies demonstrating liver toxicity in mice and pigs. The widespread occurrence of livestock poisonings by microcystins was also a consideration in the development of the guideline. Over a dozen countries have adopted the WHO guideline for microcystin-LR, such as Brazil, New Zealand, Japan, and others [8]. Other countries have established drinking water limits similar to the WHO provisional guideline. For example, Australia has established a limit of 1.3 ppb for total microcystins in drinking water. Canada has established a limit of 1.5 ppb for microcystin-LR.

Only a few countries have established drinking water standards for HAB toxins other than microcystins. For example, New Zealand has established limits for total anatoxins of 6 ppb. In addition, New Zealand has a limit of 1 µg/L for anatoxin-a, nodularin, cylindrospermopsin, and saxitoxins. Brazil has established nonenforceable recommended drinking water limits of 3 µg/L for saxitoxins and 15 µg/L for cylindrospermopsin.

HAB toxins in drinking water are not currently regulated at the federal level in the United States. Although not currently regulated, the USEPA is required to publish a list, or what is called the Contaminant Candidate List or CCL, of known or suspected compounds in drinking water that may pose a risk to the public. The USEPA is required to make a regulatory determination for at least five contaminants on the CCL every 5 years. In support of this process, the USEPA also selects a list of at least 30 unregulated contaminants for monitoring every 5 years as part of the Unregulated Contaminant Monitoring Rule (UCMR). The UCMR includes three separate lists of contaminants: List 1 of the UCMR includes contaminants selected for "assessment monitoring," which can be completed with standard analytical techniques familiar to most drinking water utilities. List 2 of the UCMR is considered a "screening survey" and includes contaminants for which specialized equipment or procedures are required; therefore, the techniques are not appropriate for every drinking water utility. List 3 is considered a "prescreen" survey as it requires analytical methods not currently available but under development.

In developing regulations, the USEPA considers health effects, feasibility of monitoring and treatment, and cost/benefits. Regulatory determinations of compounds on the CCL may be positive, negative, or no determination. A positive determination means that the USEPA will begin the process of developing a drinking water regulation for the contaminant. A negative determination indicates that the USEPA will not develop a drinking water regulation. A decision of no regulatory determination indicates that information is not sufficient to make a final determination at the time of the review.

In March 1998, the USEPA selected "cyanobacteria (blue-green algae), other freshwater algae, and their toxins" along with over 60 other chemical and microbial contaminants for the first CCL (so-called CCL1). Of the over 60 contaminants, USEPA determined that information was sufficient to make a regulatory determination for 9 of the compounds and microbial contaminants on the list. Cyanobacteria, other freshwater algae, and their toxins were not selected for further regulatory determination as part of the CCL1. Cyanobacteria, however, were placed on the UCMR List 3, indicating that further research is needed for the monitoring of these contaminants in drinking water.

In February 2005, "cyanobacteria, other freshwater algae, and their toxins" were selected for inclusion in the second CCL (CCL2). After further review of CCL2, cyanobacteria, other freshwater algae, and their toxins were not selected for regulatory determination and were not selected for further monitoring for the UCMR. In October 2009, the USEPA developed CCL3 and included three specific cyanotoxins: anatoxin-a, microcystin-LR, and cylindrospermopsin. But, again, these compounds were not selected for further regulatory review or monitoring. The draft version of CCL4 is expected to be published by USEPA in late 2014.

No drinking water standards are currently in place for HABs or HAB toxins at the state level in the United States. However, many states recognize the risk from HABs and HAB toxins and are developing programs to minimize these risks in both drinking water and recreational water. These state-level programs generally include a combination of source water protection, monitoring, event response procedures, treatment optimization or modification, and further review for possible regulatory determination. In some cases, states have established guidance values for water utilities. For example, Maryland is conducting extensive monitoring of HABs and HAB toxins through the Chesapeake Bay Monitoring Program and the recently created Coastal Bays Monitoring Program. The state of Ohio Environmental Protection Agency (OEPA) has developed a program to assist public water system operators to prevent, identify, and respond to HABs in source waters. In collaboration with the Ohio Department of Public Health and the Ohio Department of Natural Resources, the OEPA has established a HAB monitoring program and network to assess the extent of the problem and facilitate the issuing of advisories in drinking source waters. The Minnesota Department of Health has established a guidance value for the sum of all microcystin congeners in drinking water of 0.04 ppb.

4.4 Recreational Water Regulations

Like drinking water, the USEPA does not currently regulate the concentration or monitoring of HABs or HAB toxins in recreational waters. However, at least 17 states have established "guidance" or "action levels" for specific HAB toxins. States have also established programs for routine monitoring, for providing monitoring guidance, event-based response programs, and public education. For example, the state of California has established action levels of 0.8 μg/L for microcystin, 90 μg/L for anatoxin-a, and 4 μg/L for cylindrospermopsin for recreational waters. Determination of concentrations above these action levels triggers an advisory to the public. The state of Indiana has developed a slightly different approach based on different action levels for microcystin, which include

Level 1: very low/no risk < 4 μg/L microcystin-LR

Level 2: low-to-moderate risk 4 to 20 μg/L microcystin-LR

Level 3: serious risk > 20 μg/L microcystin-LR

For a level 1 advisory, the state recommends the public "use common sense practices." For a level 2 advisory, the public is advised to reduce recreational contact with water. For level 3, the public should consider avoiding all contact with contaminated water until levels of toxin decrease. Other states, such as North Carolina, Rhode Island, and Wisconsin, may issue warnings based on the visual discoloration of the water or presence of a surface scum.

TABLE 4.1

World Health Organization Guidelines
for Recreational Waters

Acute Risk	Microcystin (μg/L)	Cyanobacteria (cells/mL)
Low	4	20,000
Moderate	20	100,000
High	—	Scum

At present, there is significant variability in the type of programs at the state level and the action or guidance values established for different cyanotoxins.

WHO has established guidelines for HABs and associated toxins in recreational waters (see Table 4.1). The WHO guidelines establish levels of cyanobacteria cells and microcystin associated with different exposure risk levels. The different risk levels include "low probability of acute health effects," "moderate ... ," and "high ... ." For example, moderate risk is predicted for cyanobacterial cell counts of 100,000 cells/mL and microcystin concentrations of 20 μg/L. A number of countries, such as Australia, New Zealand, Germany, the Netherlands, and France, have also established specific action levels or guidelines for HABs and HAB toxins in recreational waters.

4.5 Nutrient Regulations

Regulations to control nutrients, including nitrogen and phosphorus, in surface water have a direct impact on the control of HABs and associated toxins. In most cases, phosphorus is the limiting nutrient in temperate freshwater lakes and nitrogen is limiting in coastal bays in estuaries [9]. Research suggested that a major factor for this observation is the greater availability of phosphorus in coastal and estuarial systems as a result of the sequestration of iron by sulfide. The salinity of seawater provides greater sulfur for the formation of iron sulfide solids, which limits the iron available for the sequestration of phosphorus. Other factors, of course, may also play a role in the observation that phosphorus is generally limiting in lakes, such as lower nitrogen fixation by cyanobacteria and higher overall rates of denitrification in coastal waters, differences in the phosphorus requirements of biota in freshwater and marine systems, and differences in nutrient processing and dynamics in these different systems.

Given the predominance of phosphorus limitation in most freshwater systems in the temperate zone, early efforts to control nutrients in lakes in the United States focused on phosphorus. The Clean Water Act of 1972, for example, established phosphorus limits for point sources such as wastewater treatment plants. Point source limits on phosphorus, along with funding available for new technology, motivated many wastewater treatment plants in the United States to upgrade to secondary treatment in the 1970s. Some wastewater treatment utilities made further improvements and upgraded to tertiary treatment, which resulted in even greater reductions in phosphorus discharges. Upgrades to wastewater treatment plants, along with bans on phosphorus in commercial products such as detergent, led to significant improvements in water quality in many surface waters and lakes, including the Great Lakes.

In 1992, the USEPA established procedures to define total maximum daily loads (TMDLs) for impaired water bodies, as authorized under section 303(d) of the Clean Water Act. States, territories, and authorized tribes were required to develop lists of impaired water bodies and develop TMDLs for these waters. The TMDL for a water body is defined as

$$TMDL = WLA + WL + MOS$$

where WLA is the waste load allocation from known point sources, WL is the waste load allocation from nonpoint sources, and MOS is a margin of safety. The TMDL is used to establish water quality-based effluent limitations for specific point sources. Point sources are issued discharge permits through the National Pollutant Discharge Elimination System (NPDES). Nonpoint sources such as agricultural sources are generally encouraged to implement best management practices (BMPs) to reduce pollutant discharges. TMDLs are established for specific water bodies and watersheds and depend on the designated use (e.g., drinking water, recreational use, aquatic life, and others) and watershed characteristics.

Confined animal feeding operations (CAFOs) are one of the largest sources of phosphorus to freshwater streams and lakes. To address this particular source of phosphorus, the USEPA is in the process of promulgating the NPDES CAFO Reporting Rule, which would require CAFOs to report basic operational information so the USEPA can best implement permitting of such facilities to protect water quality. Many states now require CAFOs to be regulated by an NPDES permit as well as require development of nutrient management plans.

Historically, only "narrative criteria" have existed to establish water quality goals related to nutrients. At the state level, efforts are currently focused on moving beyond narrative criteria for nutrient management to establish quantitative nutrient criteria for surface waters. A narrative standard uses descriptive language to describe water quality goals but does not provide numeric water quality standards that can be incorporated into the TMDL process. For example, the narrative criteria for nutrients established in the state of New York is "none in amounts that result in the growths of algae, weeds and slimes that will impair the waters for their best usages." Similarly, the narrative criteria established in New Jersey read, "Except as due to natural conditions, nutrients shall not be allowed in concentrations that cause objectionable algal densities, nuisance aquatic vegetation, or otherwise render the waters unsuitable for the designated uses." As of 2003, all the states had some type of narrative criteria for nutrients [10, 11].

In 1998, the USEPA developed the *National Strategy for the Development of Regional Nutrient Criteria* [12]. This report was followed by the National Nutrient Policy in 2001, which recommended the development of state nutrient management plans and numeric criteria for nutrients. Numeric criteria for nutrients set specific water quality targets for nitrogen and phosphorus concentrations in surface waters. The USEPA is supporting state efforts to develop numeric nutrient criteria for lakes and reservoirs, streams and rivers, and estuarine and coastal waters. To support these efforts, the USEPA established a number of "ecoregions" and provided guidance on appropriate numeric criteria. These recommendations were based on reference conditions established through analysis of various and available water quality data sets and were developed to support appropriate water quality goals based on measures of turbidity, chlorophyll-a, and dissolved oxygen.

The USEPA has established nutrient criteria, both causative and response related, for 14 distinct ecoregions across the United States (see Figure 4.1)

Draft Aggregations of Level III Ecoregions for the National Nutrient Strategy

I. Willamette and Central Valleys
II. Western Forested Mountains
III. Xeric West
IV. Great Plains Grass and Shrublands
V. South Central Cultivated Great Plains
VI. Corn Belt and Northern Great Plains
VII. Mostly Glaciated Dairy Region
VIII. Nutrient Poor Largely Glaciated Upper Midwest and Northeast
IX. Southeastern Temperate Forested Plains and Hills
X. Texas-Louisiana Coastal and Mississippi Alluvial Plains
XI. Central and Eastern Forested Uplands
XII. Southern Coastal Plain
XIII. Southern Florida Coastal Plain
XIV. Eastern Coastal Plain

FIGURE 4.1 (See color insert.)
Map of ecoregions in the United States.

Level III Ecoregions of the Conterminous United States

Map Source: USEPA, 2003

1 Coast Range	29 Central Oklahoma/Texas Plains
2 Puget Lowland	30 Edwards Plateau
3 Willamette Valley	31 Southern Texas Plains
4 Cascades	32 Texas Blackland Prairies
5 Sierra Nevada	33 East Central Texas Plains
6 Southern and Central California	34 Western Gulf Coastal Plain
Chaparral and Oak Woodlands	35 South Central Plains
7 Central California Valley	36 Ouachita Mountains
8 Southern California Mountains	37 Arkansas Valley
9 Eastern Cascades Slopes and	38 Boston Mountains
Foothills	39 Ozark Highlands
10 Columbia Plateau	40 Central Irregular Plains
11 Blue Mountains	41 Canadian Rockies
12 Snake River Plain	42 Northwestern Glaciated Plains
13 Central Basin and Range	43 Northwestern Great Plains
14 Mojave Basin and Range	44 Nebraska Sand Hills
15 Northern Rockies	45 Piedmont
16 Idaho Batholith	46 Northern Glaciated Plains
17 Middle Rockies	47 Western Corn Belt Plains
18 Wyoming Basin	48 Lake Agassiz Plain
19 Wasatch and Uinta Mountains	49 Northern Minnesota Wetlands
20 Colorado Plateaus	50 Northern Lakes and Forests
21 Southern Rockies	51 North Central Hardwood
22 Arizona/New Mexico Plateau	Forests
23 Arizona/New Mexico Mountain	52 Driftless Area
24 Chihuahuan Deserts	53 Southeastern Wisconsin Till
25 Western High Plains	Plains
26 Southwestern Tablelands	54 Central Corn Belt Plains
27 Central Great Plains	55 Eastern Corn Belt Plains
28 Flint Hills	56 Southern Michigan/Northern
	Indiana Drift Plains

57 Huron/Erie Lake Plains
58 Northeastern Highlands
59 Northeastern Coastal Zone
60 Northern Appalachian Plateau
and Uplands
61 Erie Drift Plain
62 North Central Appalachians
63 Middle Atlantic Coastal Plain
64 Northern Piedmont
65 Southeastern Plains
66 Blue Ridge
67 Ridge and Valley
68 Southwestern Appalachians
69 Central Appalachians
70 Western Allegheny Plateau
71 Interior Plateau
72 Interior River Valleys and Hills
73 Mississippi Alluvial Plain
74 Mississippi Valley Loess Plains
75 Southern Coastal Plain
76 Southern Florida Coastal Plain
77 North Cascades
78 Klamath Mountains
79 Madrean Archipelago
80 Northern Basin and Range
81 Sonoran Basin and Range
82 Laurentian Plains and Hills
83 Eastern Great Lakes and Hudson
Lowlands
84 Atlantic Coastal Pine Barrens

FIGURE 4.2 (See color insert.)
Level III ecoregions of the United States. (From USEPA at http://www.epa.gov/wed/pages/ecoregions/level_iii_iv.htm)

based on the aggregation of finer spatial resolution data in "level 3" sub-ecoregions (see Figure 4.2). The causative criteria, or criteria that lead to water quality impacts, include total phosphorus and nitrogen. Response variables are measures of nutrient impacts and include chlorophyll-a and secchi depth. For example, greater formation of algal blooms will generally

TABLE 4.2

Aggregate Ecoregion Nutrient Criteria for US Lakes and Reservoirs

Ecoregion	P (μg/L)	N (mg/L)	Chlorophyll-a (μg/L)	Secchi (m)
I	—	—	—	—
II	8.8	0.1	1.9	4.5
III	17	0.40	3.4	2.7
IV	20	0.44	2	2
V	33	0.56	2.3	1.3
VI	37.5	0.781	8.59	1.356
VII	14.75	0.66	2.63	3.33
VIII	8.0	0.24	2.43	4.93
IX	20	0.36	4.93	1.53
X	—	—	—	—
XI	8	0.46	2.79	2.83
XII	10.0	0.52	2.6	2.1
XIII	17.5	1.27	12.35	0.79
XIV	8	0.32	2.9	4.5

result in higher levels of chlorophyll-a and less water transparency as measured by secchi depth. States are encouraged to use the USEPA ecoregion nutrient criteria as a starting point in developing criteria for specific water bodies. A summary of nutrient criteria for lakes and reservoirs in the different ecoregions of the United States is shown in Table 4.2. Numeric criteria for total phosphorus for the different ecoregions vary by nearly a factor of 5, from a low of 8 μg/L in regions VIII, XI, and XIV to a high value 37.5 μg/L in region VI. For total nitrogen, levels vary over one order of magnitude, from 0.1 mg/L in ecoregion II to 1.27 mg/L in region XIII. Criteria for the response variable chlorophyll-a range from 1.9 μg/L in ecoregion II to 12.35 μg/L in region XIII. Comparison of the criteria for nitrogen and chlorophyll-a shows similar minimum and maximum values (regions II and XIII, respectively) and some correlation ($r^2 = 0.71$). Criteria for secchi depth vary from a low value of 0.79 m in region XIII to a high value of 4.93 m in region VIII. The variation in nutrient criteria observed for the different ecoregions reflects spatial variations in climate, geology, and soil types across the United States.

States vary in terms of the extent to which numeric nutrient criteria have been established or are in development. Over 20 states have not developed any state-wide nutrient criteria or criteria for individual water bodies. About a third of the states have developed partial numeric nutrient criteria, such as a phosphorus criterion for a particular water body. Three states (Rhode Island, Minnesota, and West Virginia) have established state-wide numeric criteria for phosphorus for lakes and reservoirs. Wisconsin, New Jersey, and Puerto Rico have established state-wide numeric phosphorus criteria for both lakes/reservoirs and rivers/streams. Hawaii, Guam, and the Commonwealth of Northern Marianas have developed state-wide criteria for nitrogen and phosphorus for all major and relevant water types. As an example, the state of New Jersey has established a state-wide phosphorus criterion of 0.05 mg/L for all lakes, ponds, reservoirs, and tributaries to such bodies, which is significantly higher than the ecoregion value. Wisconsin, on the other hand, has established a numeric criterion for phosphorus of 12 μg/L for lakes and reservoirs, which is within the range of phosphorus concentrations identified for the various ecoregions in the state.

The establishment of numeric nutrient criteria facilitates the development of a TMDL process to manage nutrient loads. However, only a handful of states have utilized the TMDL process to establish nutrient management programs. A few examples of nutrient TMDLs approved by the USEPA include Beaver Creek Watershed in Utah and Lake Hiddenwood, South Dakota.

Another critical piece of legislation affecting nutrients in water bodies in the United States is the Agricultural Adjustment Act, or so-called Farm Bill. The first Farm Bill was enacted in 1933 as part of the New Deal and was passed to buoy agriculture and farming during the Great Depression. In the 1980s, the Farm Bill contained, for the first time, provisions for soil conservation. Since the 1980s, the Farm Bill has evolved into one of the most significant pieces of environmental legislation in the United States specifically related to nutrients in rivers, streams, lakes, and reservoirs. In the most recently proposed Farm Bill, funds would be provided for soil conservation, implementation of BMPs, and a host of other activities aimed at reducing the environmental impact of agriculture on freshwater in the United States. Section 319 of the Clean Water Act also provides funds for BMPs.

Although most states are focusing on establishing phosphorus limits for freshwater systems, current research suggests the control of both nitrogen and phosphorus may be required in some nutrient-impacted water bodies to bring about significant water quality improvements [13]. In some impacted water bodies, for example, sediments serve as a reservoir of phosphorus. Cyanobacteria, such as *Microcystis* sp., can migrate vertically to access phosphorus in lake sediments in the absence of other sources of phosphorus. In such cases, careful control of both nitrogen and phosphorus may be required to bring about significant water quality improvements.

4.6 Case Study: Grand Lake–St. Marys

A case study is useful to consider how nutrient criteria and TMDLs are used to manage HABs and associated toxins. In the state of Ohio, a TMDL process was recently implemented for Grand Lake–St. Marys, a water body in the state severely impacted by nutrient pollution. Grand Lake–St. Marys is located in western central Ohio, near the border of Indiana. It was once the largest man-made lake in the United States, constructed in the early 1800s to serve as a reservoir for the Miami and Erie Canals. Grand Lake–St. Marys covers over 13,000 acres but averages only 5–7 feet in depth. As a result, the lake is prone to algal blooms. Historically, the lake has provided a host of recreational opportunities, such as fishing, swimming, and boating. The lake also serves as a source of drinking water for the city of Celina, Ohio. The city of Celina has a population of approximately 10,000 people.

In 2009, officials from the OEPA, as part of a USEPA study, discovered the presence of the cyanobacteria *Microcystis* as well as *Aphanizomenon*. Since 2009, blooms of harmful algae have been observed each year in Grand Lake–St. Marys. In 2010, microcystin levels of 2100 µg/L were measured in Grand Lake–St. Marys. Concentrations of cylindrospermopsin have been measured as high as 8 µg/L. As a result, the recreational value of the lake, and the positive economic impact in the region, has been severely reduced.

Grand Lake–St. Marys is located in Ecoregion VI of the Corn Belt and northern Great Plains. Within ecoregion VI, Grand Lake–St. Marys lies within the eastern Corn Belt plains (number 55) level III subregion. The Grand Lake–St. Mary watershed is heavily dominated by agriculture, with 73% of the total land area dominated by row crop agriculture. The remaining land area consists of open water (12%), land used for pasture and hay (8%), and various other minor uses. The watershed is roughly 171 square miles in area with a population of approximately 25,000 people. The watershed is also home to an estimated 295,400 "animal units" of cows, poultry, and hogs. There are nine major tributaries in the watershed that feed into Grand Lake–St. Marys.

To improve water quality in Grand Lake–St. Marys, the state of Ohio established specific numeric criteria for both nitrogen and phosphorus for the watershed. Numeric criteria were established for headwaters, wadeable waters, and small rivers. For nitrogen, target values varied from 1.0 mg/L nitrate nitrogen for headwaters to 1.5 mg/L for small rivers. Target total phosphorus concentrations ranged from 0.08 mg/L for headwaters to 0.17 mg/L for small rivers. These values for nitrogen and phosphorus were established to protect aquatic life and were based on state-wide historical data sets. The target values established for nitrogen and phosphorus for Grand Lake–St. Marys are similar, but slightly higher in some cases, to values established for USEPA ecoregion VI and subregion 55. For example, the 25th percentile values for nitrogen and phosphorus in ecoregion VI are 0.633 mg/L and 76.25 µg/L, respectively. For USEPA subregion 55, the values for nitrogen and phosphorus are 1.6 mg/L and 62.5 µg/L, respectively.

Target nutrient concentrations, historical nutrient stream data, and flow duration curves were used to establish load duration curves for nitrogen and phosphorus. Target nutrient concentrations were based on the numeric criteria discussed. Historic nutrient stream data were based on historical data sets for roughly a 10-year period. Stream flow data was obtained from US Geological Survey stream gauge stations. Continuous flow data were not available; therefore, flow duration curves were extrapolated based on discrete stream flow data. The load duration curves were estimated based on the average daily stream flow and the average nutrient concentration. Finally, TMDL values were estimated based on existing known point sources, target values, and a margin of safety. The MOS was chosen as 5%. TMDL values were calculated for each major tributary to Grand Lake–St. Marys for a variety of flow conditions, including high flow, moist conditions, mid-range flow, dry conditions, and low flows. The exact flow for each flow condition varied depending on the tributary.

Of the nine major tributaries to Grand Lake–St. Marys, Beaver Creek had the largest observed phosphorus load, with a high-flow loading of 375 kg/day. Based on the nutrient criteria values, the TMDL for Beaver Creek was determined to be 22 kg/day. As pointed out previously, the TMDL consists of LA, WLA, and MOS. For Beaver Creek, the LA, WLA, and MOS were determined to be 19, 2, and 1 kg/day, respectively. To achieve the TMDL target, a reduction in phosphorus loading of 94%, from 375 kg/day to 22 kg/day, would be required. Under dry conditions, the current phosphorus loading to Grand Lake–St. Marys from Beaver Creek was determined to be 13.5 kg/day, and a target TMDL value of 1.1 kg/day was established to meet water quality objectives. The target TMDL included a WLA of 1.05 kg/day and an MOS of 0.05 kg/day. The LA under dry conditions was estimated as zero. To meet the TMDL goal under dry conditions, a reduction in phosphorus loading of 13.5 kg/day to 1.1 kg/day would require a 92% reduction in phosphorus loading. Similar reductions were established for other major tributaries to Grand Lake–St. Marys.

Examining the TMDL reductions needed for different flow conditions showed that different activities need to be targeted under different flow conditions. For example, under high-flow conditions, the major sources of contaminants are combined sewer overflows, storm water, and agricultural field drainage. During low-flow periods, point sources, livestock, and on-site wastewater treatment systems are major contributors to contamination by phosphorus.

Finally, the TMDL targets for individual tributaries were used as input data to estimate the impact of load reductions on nutrient levels in Grand Lake–St. Marys. To accomplish this, the BATHTUB model, developed by the US Army Corps of Engineers (USACE), was used. The BATHTUB model performs water and nutrient balances and considers advection, diffusion, and nutrient sedimentation. Results from the BATHTUB model indicated that achieving TMDL reductions would improve water quality in the lake. Based on the TMDL targets, nitrogen concentration in the lake would decrease by 50% and phosphorus would decrease by 59%. As result, the BATHTUB model predicted that chlorophyll-a concentrations would decrease from 350 to 85 μg/L, while secchi depths would improve 250%. Although the TMDL targets would bring about significant water quality improvements to Grand Lake–St. Marys, the resulting nitrogen and phosphorus levels would still be significantly above USEPA established ecoregion values. For example, the predicted nitrogen concentration in Grand Lake–St. Marys based on the established TMDL reductions in tributaries would be 2.4 mg/L; the target value for lakes in ecoregion 55 is 0.782 mg/L. Similarly, the predicted phosphorus concentration in Grand Lake–St. Marys based on the TMDL reductions was 100 µg/L compared to an ecoregion target value of 35 µg/L.

The TMDL process was used to develop a number of priority strategies to meet water quality goals for Grand Lake–St. Marys. Through the TMDL process, the primary sources of nutrient pollution to Grand Lake–St. Marys were identified as livestock manure drainage, runoff from row agricultural operations, channel degradation, and point sources. To address point sources, it was recommended to modify new and existing permits to meet TMDL targets for WLA. It was estimated that it would cost roughly $475/day for a 1-million-gallon/day wastewater treatment plant to implement new processes to control phosphorus effluent to meet the TMDL targets. To address agricultural, nonpoint sources of pollution, the TMDL process identified a number of BMPs, including adjusting crop rotations, vegetative filter strips, more efficient fertilizer application, and development of nutrient management plans. The Farm Bill was identified as a potential source of cost sharing and financial support to implement some of these BMPs.

References

1. D.M. Anderson, S.B. Galloway, and J.D. Joseph, *Marine Biotoxins and Harmful Algae: A National Plan*, WHOI Technical Report-93-02. Woods Hole Oceanographic Institution, Woods Hole, MA, 1993, p. 44.
2. NOAA. Available at HYPERLINK "http://ecosystems.noaa.gov/statutory_mandates.htm" http://ecosystems.noaa.gov/statutory_mandates.htm
3. E.B. Jewett, C.B. Lopez, Q. Dortch, S.M. Etheridge, and L.C. Backer, *Harmful Algal Bloom Management and Response: Assessment and Plan*. Interagency Working Group on Harmful Algal Blooms and Human Health of the Joint Subcommittee on Ocean Science and Technology, Washington, DC, 2008.
4. C.B. Lopez, Q. Dortch, E.B. Jewett, and D. Garrison, *Scientific Assessment of Marine Harmful Algal Blooms*. Interagency Working Group on Harmful Algal Blooms and Human Health of the Joint Subcommittee on Ocean Science and Technology, Washington, DC, 2008.
5. C.B. Lopez, E.B. Jewett, Q. Dortch, B.T. Walton, and H.K. Hudnell, *Scientific Assessment of Freshwater Harmful Algal Blooms*. Interagency Working Group on Harmful Algal Blooms and Human Health of the Joint Subcommittee on Ocean Science and Technology, Washington, DC, 2008.

6. Committee on Environment and Natural Resources, *Scientific Assessment of Hypoxia in US Coastal Waters*. Interagency Working Group on Harmful Algal Blooms and Human Health of the Joint Subcommittee on Ocean Science and Technology, Washington, DC, 2010.
7. World Health Organization, *Guidelines for Drinking-Water Quality*, 2nd ed. *Addendum to Vol. 2. Health Criteria and Other Supporting Information*. World Health Organization, Geneva, Switzerland, 1998.
8. M. Burch, Effective doses, guidelines and regulations, in: K.H. Hudnell (Ed.), *Cyanobacterial Harmful Algal Blooms: State of the Science and Research Needs*. New York: Springer, 2008, p. 833.
9. S. Blomqvist, A. Gunnars, and R. Elmgren, Why the limiting nutrient differs between temperate coastal seas and freshwater lakes: a matter of salt. *Limnology and Oceanography*, 49 (2004) 2236–2241.
10. New York State Department of Environmental Conservation. Available at http://www.dec.ny.gov/chemical/77704.html.
11. New Jersey Nutrient Criteria Enhancement Plan, New Jersey Department of Environmental Protection, page 4, April, 2009.
12. US Environmental Protection Agency, *National Strategy for the Development of Regional Nutrient Criteria*. United States Environmental Protection Agency, Washington, DC, 1998.
13. D.J. Conley, H.W. Pearl, R.W. Howarth, D.F. Boesch, S.P. Seitzinger, K.E. Havens, C. Lancelot, and G.E. Likens, Controlling eutrophication: nitrogen and phosphorus. *Science*, 323 (2009) 1014–1015.

FIGURE 1.1

Satellite image of the western basin of Lake Erie in summer 2011. The image shows a massive harmful algae bloom reaching from the southern to the northern shore. (Image courtesy of NOAA.)

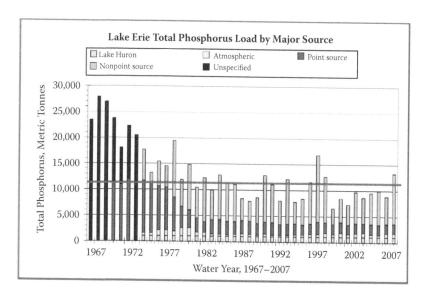

FIGURE 2.1

Annual loading of phosphorus to Lake Erie by major source. Reductions in point loads significantly reduced phosphorus inputs by the 1980s. (From the Ohio Lake Erie Phosphorus Task Force.)

Experimental Lake Erie Harmful Algal Bloom Bulletin

National Centers for Coastal Ocean Science and Great Lakes Environmental Research Laboratory
17 September 2012; Bulletin 16

The bloom has weakened in Maumee Bay since last week's bulletin. The water temperature is beginning to cool which could further decrease growth. Last Thursday, a Public Health Advisory was posted for Maumee Bay State Park beach. We forecast for a S-SE transport.
- Dupuy, Wynne, Briggs, Stumpf

Figure 1. MODIS Cyanobacterial Index from 16 September 2012.

Figure 2. Nowcast position of bloom for 17 September 2012 using GLCFS modeled currents to move the bloom from the 16 September 2012 image.

Figure 3. Forecast position of bloom for 20 September 2012 using GLCFS modeled currents to move the bloom from the 16 September 2012 image.

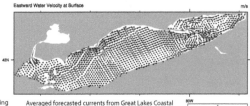

Averaged forecasted currents from Great Lakes Coastal Forecasting System over the next 72 hours.

To subscribe to this bulletin, go to :
http://www.glerl.noaa.gov/res/Centers/HABS/lake_erie_hab/signup.php

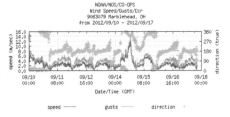

Air and Water Temperature from Marblehead, OH. From: NOAA/Center for Operational Oceanographic Products and Services (CO-OPS).

Wind Speed, Gusts and Direction from Marblehead, OH. From: NOAA/Center for Operational Oceanographic Products and Services (CO-OPS).

To subscribe to this bulletin, go to :
http://www.glerl.noaa.gov/res/Centers/HABS/lake_erie_hab/signup.php

FIGURE 2.3

Example of NOAA HAB bulletin for Lake Erie.

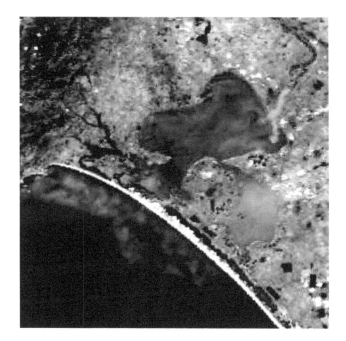

FIGURE 3.1

Algal bloom in Lake Alexandrina in southern Australia.

Draft Aggregations of Level III Ecoregions
for the National Nutrient Strategy

I. Willamette and Central Valleys
II. Western Forested Mountains
III. Xeric West
IV. Great Plains Grass and Shrublands
V. South Central Cultivated Great Plains
VI. Corn Belt and Northern Great Plains
VII. Mostly Glaciated Dairy Region
VIII. Nutrient Poor Largely Glaciated Upper Midwest and Northeast
IX. Southeastern Temperate Forested Plains and Hills
X. Texas-Louisiana Coastal and Mississippi Alluvial Plains
XI. Central and Eastern Forested Uplands
XII. Southern Coastal Plain
XIII. Southern Florida Coastal Plain
XIV. Eastern Coastal Plain

FIGURE 4.1

Map of ecoregions in the United States.

Level III Ecoregions of the Conterminous United States

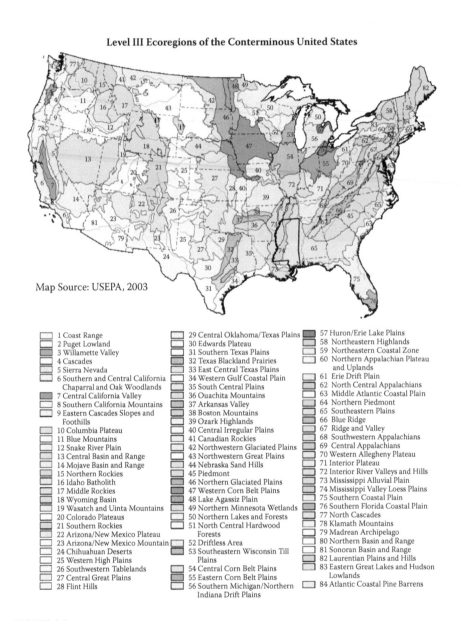

Map Source: USEPA, 2003

1 Coast Range
2 Puget Lowland
3 Willamette Valley
4 Cascades
5 Sierra Nevada
6 Southern and Central California
 Chaparral and Oak Woodlands
7 Central California Valley
8 Southern California Mountains
9 Eastern Cascades Slopes and
 Foothills
10 Columbia Plateau
11 Blue Mountains
12 Snake River Plain
13 Central Basin and Range
14 Mojave Basin and Range
15 Northern Rockies
16 Idaho Batholith
17 Middle Rockies
18 Wyoming Basin
19 Wasatch and Uinta Mountains
20 Colorado Plateaus
21 Southern Rockies
22 Arizona/New Mexico Plateau
23 Arizona/New Mexico Mountain
24 Chihuahuan Deserts
25 Western High Plains
26 Southwestern Tablelands
27 Central Great Plains
28 Flint Hills

29 Central Oklahoma/Texas Plains
30 Edwards Plateau
31 Southern Texas Plains
32 Texas Blackland Prairies
33 East Central Texas Plains
34 Western Gulf Coastal Plain
35 South Central Plains
36 Ouachita Mountains
37 Arkansas Valley
38 Boston Mountains
39 Ozark Highlands
40 Central Irregular Plains
41 Canadian Rockies
42 Northwestern Glaciated Plains
43 Northwestern Great Plains
44 Nebraska Sand Hills
45 Piedmont
46 Northern Glaciated Plains
47 Western Corn Belt Plains
48 Lake Agassiz Plain
49 Northern Minnesota Wetlands
50 Northern Lakes and Forests
51 North Central Hardwood
 Forests
52 Driftless Area
53 Southeastern Wisconsin Till
 Plains
54 Central Corn Belt Plains
55 Eastern Corn Belt Plains
56 Southern Michigan/Northern
 Indiana Drift Plains

57 Huron/Erie Lake Plains
58 Northeastern Highlands
59 Northeastern Coastal Zone
60 Northern Appalachian Plateau
 and Uplands
61 Erie Drift Plain
62 North Central Appalachians
63 Middle Atlantic Coastal Plain
64 Northern Piedmont
65 Southeastern Plains
66 Blue Ridge
67 Ridge and Valley
68 Southwestern Appalachians
69 Central Appalachians
70 Western Allegheny Plateau
71 Interior Plateau
72 Interior River Valleys and Hills
73 Mississippi Alluvial Plain
74 Mississippi Valley Loess Plains
75 Southern Coastal Plain
76 Southern Florida Coastal Plain
77 North Cascades
78 Klamath Mountains
79 Madrean Archipelago
80 Northern Basin and Range
81 Sonoran Basin and Range
82 Laurentian Plains and Hills
83 Eastern Great Lakes and Hudson
 Lowlands
84 Atlantic Coastal Pine Barrens

FIGURE 4.2
Level III ecoregions of the United States.

5

Source Water Control of Harmful Algal Blooms and Toxins

5.1 Introduction

When efforts to reduce the drivers leading to harmful algal blooms (HABs) have not taken place or are ineffective, lake managers are faced with the challenge of mitigating the impacts of HABs. One approach to mitigate the impacts of HABs on human health and the environment is to manipulate the lake or reservoir by altering the physical, chemical, or biological environment such that HABs are less likely to form. One way is to add a coagulant or adsorbent to the lake to reduce lake nutrient concentrations or more effectively settle out HAB biomass. Aluminum sulfate, or "alum," is a metal salt coagulant long used for the coagulation and removal of turbidity in conventional drinking water treatment plants. Alum also has a long history of use for the precipitation of phosphorus from wastewater. Alum has recently been explored as an *in situ* treatment chemical for the mitigation of HABs in inland lakes. To control HABs, alum is added to reduce phosphorus levels in the lake or reservoir through either precipitation or adsorption. Once phosphorus is associated with aluminum precipitates, the solids settle, and the phosphorus becomes "locked" in the lake sediments. The addition of alum during a bloom may also help remove HAB cells from the water column.

Other approaches for mitigating HABs *in situ* are also being considered. In some locations, mechanical mixing has been used to destratify the water column to provide greater mixing and reduce the average nutrient concentration, especially in the photic zone. If alternative low-nutrient stream flows are available, water managers may increase inputs to a lake or reservoir from these alternative sources to promote the flushing of nutrients from the system. Few surface water systems, however, have such alternate sources of water for flushing. The addition of clay to a nutrient-impacted water body may act in a similar way as aluminum sulfate by promoting the adsorption and sedimentation of nutrients as well as aggregating biomass. For many decades, water managers have utilized various types of algicides to control blooms. In the case of HABs, however, the use of an algicide such

TABLE 5.1

HAB Mitigation Approaches in Lakes and Reservoirs

	Processes		
Natural Attenuation	Physical Manipulation	Chemical Manipulation	Biological Manipulation
Photolysis	Artificial destratification	Alum coagulation	Biocontrol
Adsorption	Aeration	PACl	Ecological engineering
Biodegradation		Ca, Fe salts	
		Clay	
		Algicides	

as copper sulfate may disrupt the integrity of cell walls and result in significant release of HAB toxins. As discussed in Chapter 1, the use of copper sulfate is suspected to be the cause of elevated levels of cylindrospermopsin in Solomon Dam, which led to the hospitalization of over 100 children on Palm Island in Queensland, Australia, in 1979. Finally, biological methods have also been explored to shift ecosystem dynamics in an attempt to inhibit HAB formation.

In this chapter, various approaches for the in situ mitigation of HABs are explored. The chapter starts with a discussion of the processes leading to the natural attenuation of HABs and HAB toxins. Then, the various *in situ* mitigation approaches introduced in this section, including coagulation, adsorption, destratification, flushing, algicides, and biological manipulation, are presented and discussed. A list of the various approaches is provided in Table 5.1.

5.2 Natural Attenuation of HABs and Toxins

In developing *in situ* mitigation approaches for HABs, especially HAB toxins, it is helpful to understand the natural processes controlling the fate and transport of these compounds in natural systems. The fate of HABs is affected by lake ecology and factors controlling bloom dynamics, which were discussed in Chapter 2. The factors that influence toxin release from HAB cells were also discussed in Chapter 2. In this section, we follow up on these processes and discuss the primary factors affecting the fate and transport of HAB toxins once released into the environment. The primary processes affecting the fate of HAB toxins in natural waters include photolysis, adsorption, and microbial degradation. Each of these processes depends on specific conditions in the lake or reservoir, the characteristics and properties of the HAB toxins, and other factors. The magnitude and relative importance of these processes will influence the fate, transport, and impact of HAB toxins on natural systems and human health.

5.2.1 Photolysis

Photolysis is an important pathway in the destruction of many HAB toxins in natural systems. With respect to HABs, the term *photolysis* refers to the chemical transformation of toxins (e.g., HAB toxins) caused by reactions with photons. Photolysis may involve a "direct" mechanism in which the photon is absorbed directly by the toxin or an "indirect" mechanism in which another compound (i.e., a photosensitizer) absorbs the photon and produces a secondary radical species (e.g., hydroxyl radical) that is responsible for the chemical decomposition of the toxin. Natural organic matter (NOM), for example, is known to participate in indirect photolysis, thereby promoting the decomposition of toxins in natural waters. Other compounds in natural waters may act to quench photolytic reactions (i.e., the compounds absorb photons and as a result reduce the potential for photolytic reactions by other chemical compounds).

The photolytic degradation rate of HAB toxins in natural waters can be slow. Wormer et al. [1], for example, examined the photodegradation of microcystins in lake water. Experiments were conducted in situ by placing microcystin extracts in Whirl-Pak™ bags at different depths in Valmayor Reservoir, Spain. These researchers found microcystins were persistent in the water column and had half-lives ranging from 6.3 to 52 days, depending on the depth at which the samples were placed in the water column and the attenuation coefficient. The natural photodegradation of microcystins decreased significantly with depth. Applying different Mylar sheets to the Whirl-Paks showed that ultraviolet (UV) A, UVB, and photosynthetically active radiation (PAR) all contributed to the degradation of microcystins. Thirumavalavan et al. [2] found that humic acid acted as a photosensitizer and increased photodegradation rates of microcystins in natural waters. Given the variable and complex nature of natural OM, it can be difficult to predict the extent of photodegradation of HAB toxins in natural waters.

Suspended sediment may influence the photodegradation of algal toxins as well by reducing light transparency in the water column and by adsorbing toxins and reducing photolytic chemical reactions. Thirumavalavan et al. [2], however, found that suspended sediment had little effect on the photodegradation of microcystins.

The photochemical degradation of cylindrospermopsin appears to be slower than for microcystin and proceeds via an indirect mechanism. The stability of cylindrospermopsin, for example, was examined in both high-purity water and natural water samples [3]. Experiments showed little or no photochemical destruction of cylindrospermopsin in Milli-Q water, but degradation did occur in the presence of photosensitizers. Song et al. [4] also examined the photodegradation of cylindrospermopsin exposed to solar radiation in laboratory experiments. They found that NOM-mediated reactions accounted for approximately 70% of the degradation of cylindrospermopsin. Further, the breakdown of cylindrospermopsin was caused primarily by hydroxyl radical attack at the uracil functional group rather than attack by singlet oxygen or other radical species.

Anatoxin-a has a similarly slow degradation rate by natural photolysis, with half-lives on the order of days to weeks. Kaminski et al. [5] examined the degradation of anatoxin-a under simulated sunlight conditions and found photolysis was greatest at pH values of 8–10. For example, about 50% of the original concentration of anatoxin-a was present after 60 days at pH 9.5; only 3.5% of the toxin was lost over this same time interval at pH 3.5. At neutral conditions, about 60% of the toxin remained after 60 days.

5.2.2 Adsorption

Adsorption is an important process controlling the fate and transport of contaminants in natural aquatic systems, both groundwater and surface water. Adsorption influences the sedimentation rate of contaminants and bioavailability. The adsorption of cyanotoxins in natural waters is controlled by the specific characteristics of the toxins (e.g., hydrophobicity, charge density, chemical functionality), properties of the sediment material, and water chemistry. In general, significant adsorption of HAB toxins is expected to occur, and this adsorption will influence other removal processes, such as photolysis and biodegradation.

Adsorption of microcystin is the most well studied and has been shown to occur on a variety of particle types, including clays, iron oxides, and natural suspended sediment material. Thirumavalavan et al. [2], for example, examined the adsorption of microcystin to suspended sediment from reservoirs and rivers in Taiwan. Based on fitting the data to the Langmuir equation, they found that the maximum adsorbed amount q_{max} was 11.82 µg/g. Liu et al. also found high levels of adsorption of microcystin to lake sediment material [6]. Morris et al. found significant (>80%) adsorption of microcystin on clay particles, specifically montmorillonite and kaolinite [7]. The adsorption of microcystin-LR on iron oxide was examined by Lee and Walker and was found to be pH dependent. The adsorption of microcystin-LR increased with decreasing pH because of more favorable electrostatic interactions between the negatively charged microcystin and the pH-dependent charge of iron oxide [8].

Wu et al. [9] carried out adsorption experiments with microcystin on 15 different sediments and elucidated how organic matter (OM) affects the magnitude and mechanism of adsorption. In general, natural OM may compete with other contaminants for surface sites on the adsorbent particles or facilitate adsorption through OM-contaminant interactions. Wu et al. found that at low sediment OM concentrations (<8%), the adsorption of microcystin decreased with increasing OM because of competition between the two for adsorption sites on the sediment surface. At higher levels of OM, on the other hand, direct interactions between microcystin and OM were important in controlling adsorption, and the extent of adsorption increased with increasing OM. Wu et al. also found that the adsorption was pH dependent with increasing adsorption with decreasing pH.

Adsorption of cylindrospermopsin to natural particles has also been shown to occur and will have an impact on the fate and transport of this toxin in the environment. Kliztke et al. [10], for example, examined the adsorption of cylindrospermopsin to a variety of natural sediment types, including silts, sands, and clays. They found little adsorption of this toxin to silts and clays, but significant adsorption was observed for "clayey sand" and an "organic mud." Klitzke et al. speculated that the lack of adsorption of cylindrospermopsin to silts and sands was caused by electrostatic repulsion between the particles and the toxin. The greater adsorption of cylindrospermopsin on the organic sediment was suggested to be facilitated by hydrogen bonds between the toxin and humic compounds in the sediment. The adsorption on the clayey soil was believed to occur because of the zwitterionic nature of the clayey particles, which facilitated attractive electrostatic interactions.

Klitzke et al. [10] also examined the adsorption of anatoxin-a and found much stronger adsorption on sediment compared to cylindrospermopsin and microcystins. They suggested that the generally stronger adsorption of anatoxin-a, compared to these other toxins, is because anatoxin-a exists as a cation under typical environmental conditions, and cylindrospermopsin and microcystins are anions. As most natural particles are negatively charged under environmentally relevant conditions, the attractive electrostatic interactions between positive anatoxin molecules and the negatively charged particle surfaces facilitate adsorption. Klitzke et al. also suggested that the adsorption of anatoxin-a to sediment particles occurred via an ion exchange mechanism and was reversible, As a result, the adsorption is affected by ionic strength and the presence of other cation molecules.

5.2.3 Biodegradation

Biological processes can play an important role in controlling the fate and transport of HAB toxins in freshwater lakes and reservoirs. Biodegradation is a complex process. It is important to understand the significance of biodegradation in destroying HAB toxins under different conditions that may occur in a lake or reservoir. A number of questions generally arise in assessing the nature and extent of biodegradation in a particular environment. What is the relative extent of degradation for different HAB toxins? How quickly does biodegradation progress? What degradation products are formed, and what are the toxicities of these products? Does biodegradation occur in the water column, in lake sediments, or both? Is biodegradation favored under oxic or anoxic conditions? What organisms carry out the degradation process? What are the genes or gene clusters involved, and what are the biochemical degradation pathways? How does adsorption to sediment have an impact on the biodegradation process? Answers to these questions are beginning to be answered and will enable a better understanding of the role of biodegradation in controlling the fate and transport of HAB toxins in lakes and reservoirs. However, we have only a rudimentary understanding of the biodegradation process at present.

Research has shown that microcystins can be degraded biologically in natural waters. Studies have identified the types of microorganisms capable of carrying out the degradation, biochemical pathways, by-products, and gene clusters involved. In one of the early studies, Jones et al. [11], for example, demonstrated the biodegradation of microcystin-LR and microcystin-RR in laboratory experiments using both pure strains and mixed cultures. Jones et al. found that a lag phase of 2–8 days occurred prior to the initiation of biodegradation, at least on first exposure to the toxin. Other studies have shown little or no lag phase occurs prior to degradation for microcystins. For example, Edwards et al. [12] showed that microcystin-LR is readily biodegraded with typical half-lives on the order of 4–18 days. Microcystin degradation occurs in both oxic and anoxic environments, with greater degradation under anoxic conditions in the presence of nitrate [13].

Jones and others have identified the bacterium *Sphingomonas* as a microorganism capable of degrading microcystin [14]. The enzymatic degradation of microcystin occurs via a three-step series of reactions [15]. The first step in the reaction occurs via the so-called hydrolytic enzyme microcystinase, or MlrA. The enzyme MlrA catalyzes the ring opening of microcystin-LR at the Adda-arg peptide bond. Additional work suggests that this degradation product is significantly less toxic than the parent compound. The second and third enzymes in the degradation pathway (i.e., MlrB and MlrC) catalyze the additional breakdown of the molecule into a tetrapeptide and other smaller, undefined peptides and amino acids. The gene cluster that encodes for these three enzymes consists of four genes (i.e., mlrA, mlrB, mlrc, and mlrC). Edwards et al. [12] have suggested that the ability to degrade microcystins may be more widespread based on laboratory studies with multiple water samples and identification of different degradation pathways. In fact, Manage et al. [16] examined 31 different freshwater bacterial isolates and determined that 10 of them possessed the ability to degrade microcystin-LR. The isolates capable of degrading microcystin-LR belonged to *Arthrobacter*, *Brevibacterium*, and *Rhodococcus* and were the first taxa not belonging to proteobacteria identified to degrade microcystin-LR. Further study has identified additional organisms capable of degrading microcystin (for reviews, see [17, 18]).

Unlike microcystin, cylindrospermopsin appears to be more resistant to biological degradation. Wormer et al. [19], for example, found little biodegradation of cylindrospermopsin by two natural biological communities over a 40-day period. Although Smith et al. [20] showed that cylindrospermopsin could be degraded in natural systems, preexposure to the toxins was required, and there was a significant lag time before biodegradation took place. No specific organisms that can carry out this degradation have yet been identified. Less is known about the biodegradation of anatoxin-a in freshwater systems. Rapala et al. [21] showed that biodegradation of anatoxin-a occurs in natural systems but is less when the toxin is adsorbed to sediment.

5.3 In Situ Mitigation of HABs and HAB Toxins

Although significant degradation or elimination of HAB toxins occurs in natural systems via photolysis, biodegradation, and sorption, the timescale of many of these processes is on the order of days to weeks. As a result, HAB toxins may persist in the water column even after the bloom has subsided. Therefore, in many locations, water managers are faced with the challenge of mitigating the effects of HABs and HAB toxins in situ to protect human health. A number of field-based approaches have been tried to mitigate the effects of HABs, such as aeration, artificial destratification, and other techniques. In this section, the more typical in situ approaches for the mitigation of HABs and HAB toxins are examined and explored.

5.3.1 Artificial Destratification and Aeration

Artificial destratification and aeration have long been used as a strategy to improve water quality and mitigate the effects of eutrophication in lakes and reservoirs [22]. The impact of artificial destratification for the mitigation of HABs, however, is less understood. The most common approaches for the artificial destratification and aeration of lakes and reservoirs include direct injection of air or oxygen using diffusers, paddle wheel mixers, solar-powered circulation (SPC), surface sprays, aerating weirs, and others. The main goals of artificial destratification and aeration systems are to redistribute nutrients to reduce the potential for HABs, increase dissolved oxygen to mitigate the effects of hypoxia caused by algal blooms, and increase the oxygen concentration of bottom sediments to reduce nutrient release.

Diffused aeration is one of the most common approaches for destratifying the water column and increasing oxygen levels. Diffused aeration can be accomplished using either diffusion tubes or disk systems. In most installations, a compressor near the shore of the lake or reservoir feeds air to a tube or disk diffuser, typically located near the deepest part of the water body. The rising air bubbles move cold water from the hypolimnion up to the surface of the lake or reservoir, where it mixes with oxygen-rich warmer water of the epilimnion. The resulting mixing eventually destratifies the metalimnion. Although some diffusion of oxygen occurs directly from the rising bubbles, it is generally believed that most of the aeration occurs because of

FIGURE 5.1
Typical diffused aeration system consisting of a disk diffuser, aeration compressor, and solar power unit. (From Eagle One Golf Products.)

the increased mixing in the epilimnion and aeration from the atmosphere. The oxygen transfer efficiency from the bubbles rising into the water column is generally low. For example, the oxygen transfer efficiency for fine-bubble diffusers is on the order of 20%. For coarse-bubble diffusers, oxygen transfer efficiency is less than 10%. Numerous proprietary systems are available for diffused aeration of reservoirs and lakes (e.g., see Figure 5.1).

Other destratification systems also can be used to mitigate the potential for freshwater HABs. A paddle wheel aerator (see Figure 5.2), for example, floats on top of the reservoir or lake and mechanically mixes the water body. Because the effective distance of the paddle wheel is near the water surface, this approach is only appropriate for shallow ponds or reservoirs. Paddle wheels are more typical for maintaining oxygen levels in fishponds and other aquaculture systems. Other possible systems include fountains, aeration weirs, and other types of artificial waterfalls.

FIGURE 5.2
Typical paddle wheel aerator. (From Wikipedia Commons.)

The research on the impacts of diffused aeration on HABs is mixed. In one of the first studies, Reynolds et al. [23] examined the predominance of *Anabaena* versus diatoms in a limnetic enclosure study. They found that *Anabaena* was favored under stagnant flow conditions, but under mixing conditions, diatoms dominated. Visser et al. [24] also found that artificial destratification reduced HABs. In their study, they examined the effect of diffused aeration on the predominance of *Microcystis* in Lake Nieuwe Meer, the Netherlands, over a 2-year period. They found increased mixing of the water column favored flagellates, green algae, and diatoms over *Microcystis*. It was speculated that the increased turbulence in the water column negated the competitive advantage of the buoyant *Microcystis* during conditions that are more stagnant. Apparently, the turbulence allows for other taxa to out-compete *Microcystis* because of the greater mixing. Follow-up studies have demonstrated the long-term (7-year) success of diffused aeration on mitigating *Microcystis* blooms in Lake Nieuwe Meer [25].

In a recent study, artificial aeration was used successfully to reduce the magnitude and extent of a *Microcystis* bloom in a small urban lake (Tegeler See) in Berlin, Germany [26]. The Tegeler See is an important source reservoir because it provides approximately 20% of the drinking water for the city of Berlin. The aeration system consisted of compressed air fed to a disk diffuser and mixing chamber/distributor. The study examined both continuous aeration during the summer and surge aeration. It was found that continuous aeration over the summer resulted in conditions more favorable for *Microcystis* growth; surge aeration allowed for the dinoflagellate *Ceratium* sp. to outcompete *Microcystis*. Therefore, surge aeration was the preferred approach for mitigating HABs for this particular water reservoir.

An additional study examined the long-term use of artificial destratification in the North Pine Dam in Brisbane, Australia, over a period of 18 years [27]. The North Pine Dam is an important source of drinking water for the city of Brisbane and is affected by blooms of *Cylindrospermopsis*. The aeration system was a typical bubble diffuser system. Over the 18-year period, the aeration system successfully reduced chlorophyll a concentrations but did not reduce the prevalence of *Cylindrospermopsis*.

Solar-powered circulation systems are being increasingly used to prevent HABs in freshwater systems [28]. The commercially available SolarBee system is an example of an SPC. The SolarBee system consists of pontoons to support the solar cells, electric motor, and mixing apparatus (see Figure 5.3). The mixing blade induces direct flow from near the bottom of the lake or reservoir to the lake surface within a tubular intake hose. Water is discharged from the system along the lake surface. The technology was recently field tested in three nutrient-impacted reservoirs: Crystal Lake, Iowa; East Gravel Lake 4, Colorado; and Lake Palmdale, California. At all three locations, freshwater HABs were strongly suppressed, and the predominance of chlorophytes and diatoms increased because of SPC. Furthermore, suppression of freshwater HABs improved with continued use of the SPC over multiple bloom seasons. For example, cyanobacteria bloom intensity decreased in one lake by 82% the first year the SPC was used and decreased by 95% in the second year. Although some failures of SPC systems in mitigating freshwater HABs have occurred, positive results have been obtained in the majority of cases [28]. The major advantage of SPC systems is the solar power system, which eliminates the need for an electrical power system and reduces consumption of fossil fuels.

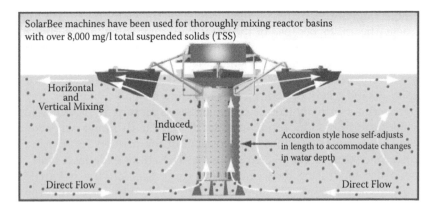

FIGURE 5.3
The SolarBee system, an example of a solar circulation system. (From Medora Corp.)

5.3.2 Alum Coagulation

Alum is a metal salt, long used for the coagulation and removal of turbidity in conventional drinking water treatment plants [29]. When alum is added to water in a conventional water treatment plant, the aluminum ions react with hydroxide in the water to form a soluble aluminum hydroxide precipitate. This solid precipitate acts as a glue to facilitate the agglomeration of fine particles and enhance their settling in sedimentation basins. In waters containing phosphorus, aluminum ions may directly react with phosphate ions to form an insoluble aluminum phosphate precipitate, which can be subsequently removed by sedimentation. Or, under certain pH conditions, phosphate ions may adsorb to aluminum hydroxide precipitates and be removed during sedimentation as an adsorbed complex. To control HABs, alum is used to reduce phosphorus levels in the lake or reservoir through either precipitation or adsorption as described previously. Once associated with aluminum precipitates, the solids settle, and the phosphorus becomes "locked" in the lake sediments. The addition of alum during a bloom may also help remove HAB cells from the water column.

Although the application of coagulants may help reduce phosphate and HAB biomass, the application of such approaches is challenging. For one, application of chemicals like alum is costly for even a moderate-size lake. To minimize cost, a good understanding of the bloom location and extent is needed. If successful, it is still largely unknown what long-term impacts may occur as a result of the application of alum in natural systems. It is not known, for example, whether the phosphate locked within the sediments can be reintroduced into the water column if sediment chemistry changes (e.g., becomes anaerobic) or significant resuspension occurs. Also, the long-term ecological impacts of alum, especially for benthic species, are not well characterized.

In Grand Lake–St. Marys, a small inland reservoir once part of the canal system in rural Ohio, HABs have had a major impact on local tourism and have also threatened drinking water. As mentioned in the case study in Chapter 4, the lake is subject to a "perfect storm" of factors that contribute to the occurrence of HABs, including shallow water depth, large nonpoint-source inputs of nutrients, and extensive channelization of the watershed to support agricultural production. In the face of this perfect storm of causes, water managers have resorted to treating Grand Lake–St. Marys with aluminum sulfate (alum) (see Figure 5.4).

Aluminum ions react with hydroxide in the water according to the following set of reactions:

$$Al_2(SO_4)_3 \rightarrow 2Al^{2+} + 3SO_4^{2-}$$

$$Al^{3+} + 3H_2O \rightarrow Al(OH)_{3(s)} + 3H^+$$

FIGURE 5.4
Liquid alum addition to Grand Lake–St. Marys. (From T. Cummins, Liquid alum project almost finished at Grand Lake St. Marys. June 22, 2011. http://www.hometownstations.com/story/14959331/liquid-alum-project-almost-finished-at-grand-lake-st-marys.)

The solid $Al(OH)_{3(s)}$ precipitate acts as a glue to facilitate the agglomeration of fine particles in water treatment plants and enhance the settling of solids in sedimentation basins.

In lakes and reservoirs affected by HABs, alum is used to remove phosphorus and settle out HAB biomass. The resulting precipitate is also expected to settle to the lake bottom and reduce the potential for the reintroduction of phosphate from lake sediments. Alum has long been used to remove phosphate in wastewater treatment processes [30]. In water containing phosphorus, aluminum ions may directly react with phosphate ions to form an aluminum phosphate precipitate, based on the following reactions:

$$Al_2(SO_4)_3 \rightarrow 2Al^{2+} + 3SO_4^{2-}$$

$$Al^{3+} + PO_4^{3-} \rightarrow AlPO_4$$

The $AlPO_4$ complex has a minimum solubility at approximately pH 6; therefore, consideration must be given to ensure lake or reservoir pH values in this range to minimize alum consumption and free aluminum. It should also be noted that aluminum ions may also react with water molecules, as described previously, which results in the production of protons. This production of protons will tend to reduce the pH of the lake or reservoir. It is important, therefore, to consider the alkalinity of the water body and factor this into calculations regarding expected changes in pH on addition of alum for phosphate removal.

The competition between the precipitation of $AlPO_4$ and $Al(OH)_3$ is important to consider for effective reduction of lake phosphorus using alum. Under equilibrium conditions, the solubility of $AlPO_4$ will be at a minimum at about pH 5.5. The minimum solubility of $Al(OH)_3$ occurs at a slightly higher pH value close to 7. Therefore, significant precipitation of $Al(OH)_3$ would be expected at pH values above 7, which would increase the dosage of alum needed for phosphate removal. The ease of settling of $AlPO_4$ is also important to consider. In the treatment of wastewater, some $Al(OH)_3$ precipitation was found beneficial for the removal of fine $AlPO_4$ particles. Therefore, the optimum dosage of alum for effective phosphate removal will be a dosage that minimizes $Al(OH)_3$ precipitation but provides sufficient aluminum hydroxide to aid in settling.

Under certain pH conditions, phosphate ions may adsorb to aluminum hydroxide solids and be removed during sedimentation as an adsorbed complex. Early work on the adsorption of phosphate to amorphous aluminum hydroxide, for example, showed that adsorption was pH dependent with increasing adsorption with increasing pH [31]. To control HABs, alum is used to reduce phosphorus levels in the lake or reservoir through either precipitation or adsorption as described previously. Once associated with aluminum precipitates, the solids settle, and the phosphorus becomes "locked" in the lake sediments. The addition of alum during a bloom may also help remove HAB cells from the water column.

Although alum may precipitate or adsorb phosphorus and lock it in bottom sediments and facilitate the removal of HAB biomass, the alum coagulation process likely does not aid in reducing the concentrations of dissolved toxins. Research examining the fate of HAB toxins in water treatment plants utilizing metal salt coagulation show little or no removal of toxin during this process. Rositano and Nicholson, for example, performed one of the first studies that showed that alum, ferric sulfate, and polyaluminum chloride (PACl) coagulants were all ineffective at removing dissolved toxins [32]. Although coagulation is sometimes used to achieve partial removal of natural OM (so-called enhanced coagulation), the affinity of natural OM for floc particles is considerably higher owing to the greater molecular weight and, in some cases, greater hydrophobicity. As a result, little or no removal of dissolved toxins should be expected when alum coagulation is used as a treatment for HAB mitigation in lakes and reservoirs.

Other "metal salt" coagulants could also potentially be used for the in situ reduction of lake phosphate levels. For example, ferric and ferrous iron react with phosphate to form an insoluble iron-phosphate precipitate according to the following reactions,

$$Fe^{3+} + PO_4^{3-} \rightarrow FePO_4$$

$$3Fe^{2+} + 2PO_4^{3-} \rightarrow Fe_3(PO_4)_2$$

Like alum, ferric and ferrous salts also react with water to produce protons that react with the natural alkalinity of the water body. The optimum pH for phosphate removal using iron-based coagulants is in the range of 4–5 and is therefore lower than the pH needed for phosphate removal using alum.

Calcium, typically in the form of lime, is known to react with phosphate, resulting in insoluble hydroxyapatite, according to

$$5CaO + 5H_2O + 3PO_4^{3-} \rightarrow Ca_5(OH)(PO_4)_3 + 9OH^-$$

In wastewater treatment, the removal of phosphate with lime generally increases with increasing pH, with removal requiring a pH of 9 or more [30].

Some research has examined the use of aluminum, iron, and calcium for the reduction of lake phosphorus concentrations. Cooke et al. found that aluminum salts were capable of long-term (5–14 years) control of phosphorus release from sediments [33]. Iron and calcium, on the other hand, were not able to provide for long-term phosphorus control. Cooke et al. also noted that aluminum toxicity is a potential problem if the pH is not kept in the range of minimum aluminum solubility, typically around 6–8. More recent research studies have also raised concern regarding toxin release and the use of alum as an in-lake treatment for preventing HABs. Laboratory and microcosm experiments, for example, suggested that alum treatment may release up to 97% of the intercellular HAB toxins [34].

The use of PACl, a polymerized form of aluminum, has been used in combination with a modified clay for the in-lake inactivation of phosphorus [35]. The authors carried out both laboratory and in-lake studies. For in-lake application of PACl and the proprietary clay, the researchers found the total phosphate (TP) decreased from 169 µg/L prior to treatment to 14 µg/L after treatment. As a result of the treatment, the lake was effectively shifted from an eutrophic/hypertrophic state to an oligo-/mesotrophic state and has remained so for 5 years following treatment.

5.3.3 Clay Flocculants

As an alternative to chemical coagulation, the use of clay has been explored for many years as a means to flocculate and settle HABs and mitigate the effect of HABs [36, 37]. In marine systems, for example, ample evidence supports the use of a variety of clays and clay-rich sediment, including montmorillonite, kaolinite, and others to flocculate HAB biomass and accelerate the

FIGURE 5.5
Application of clay particles to mitigate marine HAB. (From D.M. Anderson, *Ocean and Coastal Management*, 52 (2009) 342–347.)

sedimentation process [38]. Clay flocculants have been used successfully to mitigate a number of different classes of marine HABs, including *Aureococcus* sp., *Karenia brevis*, *Heterosigma akashiwo*, and *Pfiesteria piscicida*. An application of clay particles to control a marine HAB is shown in Figure 5.5. Although the addition of clay flocculants can improve the settling of HABs, the impact on the benthic ecosystem is less well known.

Pan et al. [39] conducted one of the most comprehensive studies examining the use of clay and clay-rich sediment for the mitigation of *Microcystis* blooms. They studied the ability of 26 different natural clays and minerals for flocculating *Microcystis*. The 26 natural clays and minerals were classified as one of three "types" depending on their treatment effectiveness after 8 hours: type 1 (>85%), type 2 (50–85%), and type 3 (<50%). The type 1 clays and minerals included talc, ferric oxide, sepiolite, ferroferric oxide, and kaolinite.

The mechanism of clay flocculation of HABs is not well understood but can be explained in terms of traditional "DLVO" theory (DLVO stands for Derjaguin and Landau, Verwey and Overbeek, the founders of the theory). The flocculation of particles (e.g., a HAB cell and a clay particle) occurs in two different steps: transport and sticking. For flocculation of these two particles to occur, the different particles must be transported into close proximity and collide. Important transport processes include diffusion, shear (via mixing), and differential sedimentation. Once particles collide, the extent of aggregation will depend on the forces between the two particles.

DLVO theory explains the flocculation of fine particles as a balance between attractive van der Waals forces and electrostatic interactions. Electrostatic interactions depend on the sign and magnitude of the charge as well as solution conditions. In most cases, clay particles and harmful algal cell surfaces are both negatively charged at environmentally relevant pH conditions. This similarity in charge results in a net repulsive interaction. In marine systems, however, the high ionic strength reduces the electrostatic repulsion between like-charged surfaces through a mechanism called "charge screening," thus allowing for aggregation. In freshwater systems, the ionic strength will be much lower; therefore, electrostatic forces between HAB cells and clay particles will more likely be repulsive.

In one of the classic studies in this area, Avnimelech et al. [40] examined the mutual flocculation of bentonite clay with a number of different algae, including *Euglena gracilis*, *Anabaena* sp., *Chlomydomonas* sp., and *Chlorella* sp. They found, as expected based on the charge screening mechanism described, that the mutual flocculation was dependent on salt concentration. The researchers identified the "critical coagulation concentration (ccc)," or concentration of salt that induced significant flocculation, for different electrolytes. At concentrations below the ccc, little or no mutual flocculation was observed; above the ccc, rapid flocculation was seen. For *Chlorella* sp., the ccc was found to be 2×10^{-2} M for NaCl and 5×10^{-4} for $CaCl_2$. The lower ccc for the divalent $CaCl_2$ salt, compared to monovalent NaCl, is expected based on charge balance considerations. Pan et al. documented the net negative charge of most natural clays and minerals as well as *Microcystis* cells [39] and suggested that flocculation in freshwater systems was therefore dependent on a mechanism other than charge screening. The sepiolite clay particles examined by Pan et al., for example, were shown in scanning electron micrographs to form a fibrous structure that acted like a "net" to effectively capture *Microcystis* cells during the mutual flocculation process.

One particular concern regarding the use of clay for the mutual flocculation of HAB cells is the potential for the resuspension of HAB-clay complexes during times of high turbulence. Beaulieu et al. [41] examined how the flow environment influenced the resuspension of *Heterocapsa triquetra* and phosphatic clay (IMC-P4) complexes in a 17-m "race track" flume. They found that the extent of resuspension, and the bed shear stress required for resuspension, depended on the time the flocs had to consolidate on the bed. Relatively low flows and bed shear rates were required for resuspension after only a few hours, and the flow rate and bed shear needed for resuspension increased even after 9 hours of consolidation. They also found that the use of PACl, a prepolymerized metal salt coagulant, increased the shear stress needed to resuspend the clay-algae flocs.

5.3.4 Algicides

Algicides have long been used by lake managers to control algal blooms in lakes, drinking water reservoirs, and small ponds. Examples of algicides include copper sulfate, Polyquat (n-alkyl-dimethyl benzyl ammonium chloride), and barley straw. Copper sulfate, however, is the most commonly used algicide in lakes and reservoirs [22]. The cupric ion (Cu^{2+}) is toxic to algae at relatively low dosages. For example, cyanobacteria can be sensitive to copper at concentrations as low as 5 µg/L. A typical application of copper sulfate in surface waters for algae control utilizes a copper concentration of 500 µg/L. It is believed that the cupric ion interacts with the cell membranes and disrupts photosynthesis [22]. Copper is consumed in the process; therefore, the effect of copper sulfate treatment is generally short-lived, on the order of a few days. This also means that the amount of copper sulfate needed can be prohibitively expensive if the bloom occurs over a large area and a prolonged time interval. It is more common to apply copper sulfate to concentrated bloom areas, such as bloom areas concentrated by wind. It has also been shown that algae can develop resistance to copper sulfate treatment, which can make future applications prohibitively expensive because of the high levels of copper needed for effective treatment [22]. For example, Garcia-Villada et al. [42] showed that repeated use of copper sulfate resulted in resistant strains of *Microcystis*.

Of greater concern, however, is the potential for toxin release when algicides are used for the control of HABs. Jones and Orr [43], for example, showed that the application of algicide for the control of *Microcystis* in Lake Centenary resulted in the release of very high levels of microcystin, on the order of 1300 and 1800 µg/L. Ross et al. [44] found that release of microcystin-LR increased 90% when Polyquat was applied and corresponded to an increase in oxidative stress, which likely played a role in the mechanism of toxin release. Algicidal bacteria (e.g., bacterium B5) have also been shown to lyse algae cells, including *Microcystis aeruginosa, Chlorella,* and *Scenedesmus,* and release toxin [45]. Algicides may also disrupt the ability of other organisms to degrade HAB toxins [20]. Given the significance of cell lysis and toxin release, the use of algicides for controlling HABs should be undertaken with great caution.

5.3.5 Other Approaches

In addition to destratification, flocculation, and algicides, a variety of other approaches are being explored for the control of HABs in freshwater systems [37]. The development of "biocontrol" or the introduction of organisms that may prevent or hinder the growth of HAB species is being explored. Ecological engineering approaches aim at fostering ecologies that are more favorable to non-bloom-forming taxa and conditions. Changing the circulation patterns of lakes and reservoirs, through channelization, may also provide some benefit in controlling HABs. Dredging bottom sediments may eliminate in-lake sources of nutrients as well as change circulation patterns. The use of dyes to shade a lake or reservoir and reduce light penetration has also been considered.

References

1. L. Wörmer, M. Huerta-Fontela, S. Cirés, D. Carrasco, and A. Quesada, Natural photodegradation of the cyanobacterial toxins microcystin and cylindrospermopsin. *Environmental Science and Technology*, 44 (2010) 3002–3007.
2. M. Thirumavalavan, Y.-L. Hu, and J.-F. Lee, Effects of humic acid and suspended soils on adsorption and photo-degradation of microcystin-LR onto samples from Taiwan reservoirs and rivers. *Journal of Hazardous Materials*, 217–218 (2012) 323–329.
3. R.K. Chiswell, G.R. Shaw, G. Eaglesham, M.J. Smith, R.L. Norris, A.A. Seawright, and M.R. Moore, Stability of cylindrospermopsin, the toxin from the cyanobacterium, *Cylindrospermopsis raciborskii*: effect of pH, temperature, and sunlight on decomposition. *Environmental Toxicology*, 14 (1999) 155–161.
4. W. Song, S. Yan, W.J. Cooper, D.D. Dionysiou, and K.E. O'Shea, Hydroxyl radical oxidation of cylindrospermopsin (cyanobacterial toxin) and its role in the photochemical transformation. *Environmental Science and Technology*, 46 (2012) 12608–12615.
5. A. Kaminski, B. Bober, Z. Lechowski, and J. Bialczyk, Determination of anatoxin-a stability under certain abiotic factors. *Harmful Algae*, 28 (2013) 83–87.
6. G. Liu, Y. Qian, S. Dai, and N. Feng, Adsorption of microcystin LR and LW on suspended particulate matter (SPM) at different pH. *Water, Air, and Soil Pollution*, 192 (2008) 67–76.
7. R.J. Morris, D.E. Williams, H.A. Luu, C.F.B. Holmes, R.J. Andersen, and S.E. Calvert, The adsorption of microcystin-LR by natural clay particles. *Toxicon*, 38 (2000) 303–308.
8. J. Lee and H.W. Walker, Adsorption of microcystin-LR onto iron oxide nanoparticles. *Colloids and Surfaces a–Physicochemical and Engineering Aspects*, 373 (2011) 94–100.
9. X. Wu, B. Xiao, R. Li, C. Wang, J. Huang, and Z. Wang, Mechanisms and factors affecting sorption of microcystins onto natural sediments. *Environmental Science and Technology*, 45 (2011) 2641–2647.
10. S. Klitzke, C. Beusch, and J. Fastner, Sorption of the cyanobacterial toxins cylindrospermopsin and anatoxin-a to sediments. *Water Research*, 45 (2011) 1338–1346.
11. B.M. Long, G.J. Jones, and P.T. Orr, Cellular microcystin content in N-limited *Microcystis aeruginosa* can be predicted from growth rate. *Applied and Environmental Microbiology*, 67 (2000) 278–283.
12. C. Edwards, D. Graham, N. Fowler, and L.A. Lawton, Biodegradation of microcystins and nodularin in freshwaters. *Chemosphere*, 73 (2008) 1315–1321.

13. T. Holst, N.O.G. Jørgensen, C. Jørgensen, and A. Ohansen, Degradation of microcystin in sediments at oxic and anoxic, denitrifying conditions. *Water Research*, 37 (2003) 4748–4760.

14. S. Imanishi, H. Kato, M. Mizuno, K. Tsuji, and K.-I. Harada, Bacterial degradation of microcystins and nodularin. *Chemical Research in Toxicology*, 18 (2005) 591–598.

15. D.G. Bourne, P. Riddles, G.J. Jones, W. Smith, and R.L. Blakeley, Characterisation of a gene cluster involved in bacterial degradation of the cyanobacterial toxin microcystin LR. *Environmental Toxicology*, 16 (2001) 523–534.

16. P.M. Manage, C. Edwards, B.K. Singh, and L.A. Lawton, Isolation and identification of novel microcystin-degrading bacteria. *Applied and Environmental Microbiology*, 75 (2009) 6924–6928.

17. L. Ho, E. Sawade, and G. Newcombe, Biological treatment options for cyanobacteria metabolite removal—a review. *Water Research*, 46 (2012) 1536–1548.

18. D. Dziga, M. Wasylewski, B. Wladyka, S. Nybom, and J. Meriluoto, Microbial degradation of microcystins. *Chemical Research in Toxicology*, 26 (2013) 841–852.

19. L. Wormer, S. Cirés, D. Carrasco, and A. Quesada, Cylindrospermopsin is not degraded by co-occurring natural bacterial communities during a 40-day study. *Harmful Algae*, 7 (2008) 206–213.

20. M.J. Smith, G.R. Shaw, G.K. Eaglesham, L. Ho, and J.D. Brookes, Elucidating the factors influencing the biodegradation of cylindrospermopsin in drinking water sources. *Environmental Toxicology*, 23 (2008) 413–421.

21. J. Rapala, K. Lahti, K. Sivonen, and S.I. Niemelä, Biodegradability and adsorption on lake sediments of cyanobacterial hepatotoxins and anatoxin-a. *Letters in Applied Microbiology*, 19 (1994) 423–428.

22. A.J. Horne and C.R. Goldman, *Limnology*. New York: McGraw-Hill, 1994.

23. C.S. Reynolds, S.W. Wiseman, B.M. Godfrey, and C. Butterwick, Some effects of artificial mixing on the dynamics of phytoplankton populations in large limnetic enclosures. *Journal of Plankton Research*, 5 (1983) 203–234.

24. P. Visser, B.A.S. Ibelings, B. Van Der Veer, J.A.N. Koedood, and R. Mur, Artificial mixing prevents nuisance blooms of the cyanobacterium *Microcystis* in Lake Nieuwe Meer, the Netherlands. *Freshwater Biology*, 36 (1996) 435–450.

25. E. Jungo, P.M. Visser, J. Stroom, and L.R. Mur, Artifical mixing to reduce growth of the blue-green alga *Microcystis* in Lake Nieuwe Meer, Amsterdam: an evaluation of 7 years of experience. *Water Science and Technology: Water Supply*, 1 (2001) 17–23.

26. K.-E. Lilndenschmidt, Controlling the growth of *Microcystis* using surged artificial aeration. *International Review of Hydrobiology*, 84 (1999) 243–254.

27. J.P. Antenucci, A. Ghadouani, M.A. Burford, and J.R. Romero, The long-term effect of artificial destratification on phytoplankton species composition in a subtropical reservoir. *Freshwater Biology*, 50 (2005) 1081–1093.

28. H.K. Hudnell, C. Jones, B. Labisi, V. Lucero, D.R. Hill, and J. Eilers, Freshwater harmful algal bloom (FHAB) suppression with solar powered circulation (SPC). *Harmful Algae*, 9 (2010) 208–217.

29. J.C. Crittenden, R.R. Trussell, D.W. Hand, K.J. Howe, and G. Tchobanoglous, *Water Treatment: Principles and Design*, 2nd ed. Hoboken, NJ: Wiley, 2005.

30. T.J. McGhee, *Water Supply and Sewerage*. New York: McGraw-Hill, 1991.

31. P.H. Hsu and D.A. Rennie, Reactions of phosphate in aluminum systems. I. Adsorption of phosphate by x-ray amorphous "aluminum hydroxide." *Canadian Journal of Soil Science*, 42 (1962) 197–209.

32. J. Rositano and B.C. Nicholson, *Water Treatment Techniques for Removal of Cyanobacterial Toxins from Water.* Australian Centre for Water Quality Research, Salisbury, South Australia, 1994.

33. G.D. Cooke, E. Welch, A. Martin, D. Fulmer, J. Hyde, and G. Schrieve, Effectiveness of Al, Ca, and Fe salts for control of internal phosphorus loading in shallow and deep lakes, in: P.C.M. Boers, T.E. Cappenberg, and W. Raaphorst (Eds.), *Proceedings of the Third International Workshop on Phosphorus in Sediments.* Dordrecht, Netherlands: Springer, 1993, pp. 323–335.

34. J. Han, B.S. Jeon, N. Futatsugi, and H.D. Park, The effect of alum coagulation for in-lake treatment of toxic *Microcystis* and other cyanobacteria related organisms in microcosm experiments. *Ecotoxicology and Environmental Safety*, 96 (2013) 17–23.

35. M. Lürling and F. van Oosterhout, Controlling eutrophication by combined bloom precipitation and sediment phosphorus inactivation. *Water Research*, 47 (2013) 6527–6537.

36. D.M. Anderson, Turning back the harmful red tide. *Nature*, 388 (1997) 513–514.

37. D.M. Anderson, Approaches to monitoring, control and management of harmful algal blooms (HABs). *Ocean and Coastal Management*, 52 (2009) 342–347.

38. M.R. Sengco and D.M. Anderson, Controlling harmful algal blooms through clay flocculation. *Journal of Eukaryotic Microbiology*, 51 (2004) 169–172.

39. G. Pan, M.-M. Zhang, H. Chen, H. Zou, and H. Yan, Removal of cyanobacterial blooms in Taihu Lake using local soils. I. Equilibrium and kinetic screening on the flocculation of *Microcystis aeruginosa* using commercially available clays and minerals. *Environmental Pollution*, 141 (2006) 195–200.

40. Y. Avnimelech, B.W. Troeger, and L.W. Reed, Mutual flocculation of algae and clay: evidence and implications. *Science*, 216 (1982) 63–65.

41. S.E. Beaulieu, M.R. Sengco, and D.M. Anderson, Using clay to control harmful algal blooms: deposition and resuspension of clay/algal flocs. *Harmful Algae*, 4 (2005) 123–138.

42. L. García-Villada, M. Rico, M.a. Altamirano, L. Sánchez-Martín, V. López-Rodas, E.-R. Costas, Occurrence of copper resistant mutants in the toxic cyanobacteria *Microcystis aeruginosa*: characterisation and future implications in the use of copper sulphate as algicide. *Water Research*, 38 (2004) 2207–2213.

43. G.J. Jones and P.T. Orr, Release and degradation of microcystin following algicide treatment of a *Microcystis aeruginosa* bloom in a recreational lake, as determined by HPLC and protein phosphatase inhibition assay. *Water Research*, 28 (1994) 871–876.

44. C. Ross, L. Santiago-Vázquez, and V. Paul, Toxin release in response to oxidative stress and programmed cell death in the cyanobacterium *Microcystis aeruginosa*. *Aquatic Toxicology*, 78 (2006) 66–73.

45. R.-m. Mu, Z.-q. Fan, H.-y. Pei, X.-l. Yuan, S.-x. Liu, and X.-r. Wang, Isolation and algae-lysing characteristics of the algicidal bacterium B5. *Journal of Environmental Sciences*, 19 (2007) 1336–1340.

6

Conventional Treatment Processes for Removal of HAB Cells and Toxins from Drinking Water

6.1 Introduction

The emergence of a harmful algal bloom (HAB) in a water supply reservoir will place a strain on the treatment abilities of a conventional drinking water treatment plant. Historically, the primary goals of a conventional drinking water treatment facility have been to improve the clarity of drinking water, reduce odors, and prevent the spread of waterborne disease. The process flow diagram for a typical water treatment plant is shown in Figure 6.1. Processes such as coagulation, flocculation, sedimentation, filtration, adsorption, and disinfection have evolved over the years to meet these primary needs. In more recent decades, these processes have been further modified and improved to remove or reduce the formation of hazardous chemicals, such as disinfection by-products, arsenic, and a host of synthetic organic compounds.

In the face of HABs, staff and operators at a water treatment facility must understand how conventional treatment processes will influence HAB cells and toxin removal. Some conventional treatment processes may remove cells or toxins under normal operating conditions; other processes may require modification to realize significant removal of cells or toxins. Perhaps of greatest concern, some processes in a conventional drinking water treatment plant can increase toxin levels in the treated water if operating conditions are not considered prior to a bloom event. Therefore, thorough understanding of how different conventional treatment processes affect HAB cell and toxin removal is critical to appropriately respond to HAB events and ensure the highest-quality drinking water for consumers.

In this chapter, the effect of conventional drinking water treatment processes on HAB cell and toxin removal is described. A number of articles have reviewed the extent of HAB cell and toxin removal during drinking water treatment [1, 2]. The focus of this chapter, then, is the description of how

83

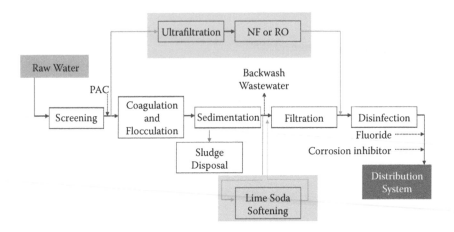

FIGURE 6.1
Process flow diagram for a drinking water treatment plant. NF = nanofiltration, PAC = powdered activated carbon, RO = reverse osmosis.

key design and operational variables influence HAB cell and toxin levels in finished drinking water.

6.2 Coagulation, Flocculation, Sedimentation, and Dissolved Air Flotation

The primary goal of the coagulation and flocculation process during drinking water treatment is to agglomerate particulate matter to facilitate solids removal during sedimentation or dissolved air flotation. A coagulant, such as "alum" or ferric chloride, is added to the water and is rapidly mixed. The coagulant serves to bind or glue the particulate matter present in the source water into flocs. Rapid mixing is applied early in this process to quickly and effectively disperse the coagulant chemicals. After rapid mixing, the water is mixed more gently through so-called flocculant mixing to facilitate the growth of large, fast-sinking flocs but at the same time prevent the breakup of previously formed flocs. Sedimentation tanks provide quiescent conditions to maximize the removal of particle flocs by settling. The clarified water leaves the sedimentation tank by passing over an outlet weir system. The "sludge" accumulates on the bottom of the sedimentation tank and is removed periodically. For HAB cells, a critical challenge is effective removal of the cells using these processes while minimizing the potential for cell disruption and the subsequent release of toxins.

The removal of HAB cells by coagulation and flocculation requires the destabilization of the cells prior to aggregation and physical separation by sedimentation or dissolved air flotation. As discussed in the previous chapter, the flocculation of HAB cells requires that cells collide and stick together. In a typical freshwater reservoir, electrostatic repulsion between HAB cells will provide a barrier to flocculation. As a result, some type of "destabilization" is needed to realize significant flocculation of HAB cells.

In general, particle destabilization during coagulation and flocculation may occur by four distinct mechanisms: (1) charge screening, which compresses the electric double layer around the particles; (2) charge neutralization; (3) adsorption of polymeric flocculants and subsequent destabilization by interparticle bridging; and (4) precipitate enmeshment, sometimes referred to as "sweep" flocculation [3]. The use of metal salts such as alum and ferric chloride during coagulation results in destabilization primarily via charge neutralization and precipitate enmeshment [3]. For coagulants such as alum, charge neutralization may occur as a result of the adsorption of positively charged hydrolysis products on the surface of HAB cells. A variety of both mononuclear and polynuclear hydrolysis products for aluminum and iron-based coagulants has been identified [3]. Because of the sensitivity of charge neutralization to the surface charge of the particles and dosage of the coagulant, sweep flocculation is usually preferred in practice. Generally, higher coagulant dosages and pH levels between 6 and 8 are needed for sweep flocculation to be operative. During sweep flocculation, particulates are removed by enmeshment in the rapidly precipitating metal hydroxide floc.

Like many algae and cyanobacteria in general, HAB cells possess a net negative surface charge or zeta potential in natural waters [4–6] because of the ionization of chemical functional groups (see Figure 6.2). The net

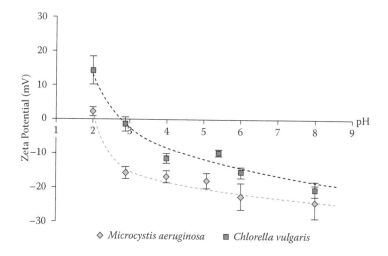

◇ *Microcystis aeruginosa*　　■ *Chlorella vulgaris*

FIGURE 6.2
Zeta potential of *Microcystis* sp. and *Chlorella* sp. as a function of pH. (From B.C. Hitzfeld, S.J. Hoger, and D.R. Dietrich, *Environmental Health Perspectives*, 108 (2000) 113–122.)

negative charge of HAB cells creates an electrostatic repulsive force between HAB cells and other negatively charged particles such as clay or sand. For example, Kwon et al. showed that the zeta potential of *Microcystis* sp. was -10.8 mV at pH 7 [5]. Hadjoudja showed that the zeta potential of *Microcystis* sp. varied from near zero at pH 2 to approximately -22 mV at pH 8 in 10 mM $NaNO_3$ solution [7]. Based on Fourier transform infrared (FTIR) spectroscopy, Hadjoudja surmised important ionizable functional groups for *Microcystis* to be carboxyl, amine, phosphoryl, and hydroxyl groups [7]. Thus, as pH increases (and proton concentrations diminish), these ionization functional groups will lose protons, resulting in greater net negative surface charge. The pH at which ionization functional groups lose protons is measured by the pK_a. Combining the FTIR results with potentiometric titrations, the mean pK_a values for the different ionization functional groups were determined as follows: $pK_1 = 3.9 \pm 0.1$ (carboxyl), $pK_2 = 7.1 \pm 0.01$ (phosphoryl), and $pK_3 = 9.0 \pm 0.3$ (amine or hydroxyl) groups. Destabilization of HAB cells requires either the neutralization of this net negative surface charge by positively charged hydrolysis products or enmeshment during sweep flocculation.

A number of studies have examined the extent of HAB cell and toxin removal in coagulation, flocculation, and sedimentation processes. In general, these processes can remove HAB cells but are ineffective at removing toxins. Teixeira et al. [8], for example, performed laboratory studies examining the removal of *Microcystis aeruginosa* from model surface waters using alum and a prepolymerized aluminum coagulant (i.e., PACl, polyaluminum chloride) and found removals of 60% to 100% (measured as chlorophyll a) depending on the source water cell concentration and coagulant dosage. It was also shown that dissolved air flotation was generally a more effective separation process than sedimentation.

Research suggested that aluminum-based coagulants are more effective at removing cyanobacterial cells than iron-based coagulants. In laboratory-scale flocculation experiments, for example, *Microcystis* sp. concentrations were reduced from 268,000 cells/mL to 64,000 cells/mL using 6 mg/L iron. Using 4 mg/L aluminum reduced the *Microcystis* sp. concentrations to 15,400 cells/mL aluminum [9]. Iron-based coagulants are generally more effective at removing natural organic matter compared to aluminum-based coagulants [3]. This suggests that the sorption of natural organic matter to floc particles may hinder the removal of HAB cells. The sorption of natural organic matter on floc particles will influence the net charge of flocs, as the natural organic matter has a net negative charge.

The treatment conditions for each coagulant described in that study, however, were not optimized in these experiments. It is generally recommended that jar testing be performed to determine the optimum type and dosage of coagulant given the complexity of source waters and the coagulation process [3]. Jar tests should be performed under conditions that match, as closely as possible, the source water quality and operational parameters of the plant. Thus, frequent jar testing will provide valuable information about the optimum coagulant dosage during the course of an HAB event. Also, process monitoring using turbidity and zeta potential can help to ensure effective coagulation of HAB cells. For example, maintaining zeta potential between -8 and $+2$ mV ensured effective coagulation in laboratory studies [4].

Rositano and Nicholson performed one of the first studies that showed that alum, ferric sulfate, and PACl coagulants were all ineffective at removing dissolved toxins [10]. These results are consistent with the fact that coagulation is not considered an effective treatment process for dissolved organic contaminants. Although coagulation is sometimes used to achieve partial removal of natural organic matter (so-called enhanced coagulation), the affinity of natural organic matter for floc particles is considerably higher owing to the increased molecular weight and, in some cases, greater hydrophobicity.

Because of the turbulence generated during coagulation, the potential exists to disrupt HAB cells and release toxins into the water. The rapid mixing stage may be short, about seconds, but is turbulent with velocity gradients of 1000/s. The main goal of rapid mixing is to disperse the coagulant and destabilize the particles. The flocculation stage is longer, approximately minutes, and mixing is gentler, with velocity gradients from 10 to 50/s. The susceptibility of HAB cells to damage will be a function of both the mixing time and the intensity.

Studies have examined the potential for cell damage and toxin release in the jar test and pilot-scale experiments. Chow et al. [11], for example, studied the potential release of toxins from *Microcystis* during simulated water treatment with aluminum sulfate. Water samples containing *Microcystis* were rapidly mixed at a velocity gradient of 480/s for 1 minute and flocculant mixed at 18/s for an additional 14 minutes. After flocculant mixing, the samples were allowed to settle for a period of 15 minutes. The results of the jar testing demonstrated no cell damage and no increase in the release of toxin from the *Microcystis* cells during the standard jar test procedure. Additional tests were carried out to examine the impact of mixing for longer times. In fact, experiments showed that no increase in microcystin release occurred even during rapid mixing at 480/s for up to 4 hours. Experiments with a pilot-scale water treatment system confirmed the jar test results and showed no damage to *Microcystis* cells and no increase in microcystin release during treatment. Similar jar test experiments by others have confirmed these results for alum and *Microcystis* [12] and shown little cell damage when PACl is used as the primary coagulant [13]. Peterson et al. [14] have shown that alum and ferric chloride also do not result in damage to *Aphanizomenon* or toxin release.

A dual oxidant-coagulant was examined for the removal of HAB toxins from drinking water. Yuan et al. [15] examined the removal of microcystin-LR using ferrate and found effective removal of this toxin over the pH range of 6–10. Fe(VI) in ferrate decomposes to Fe(III) in water; therefore, ferrate is a powerful oxidant. In fact, Yuan et al. observed 100% removal of the microcystin-LR at a dosage of 40 mg/L. The removal of microcystin was pH dependent, with greater than 80% removal observed over the pH range of

6–10. Once oxidized to Fe(III), the iron precipitates as a hydroxide, thereby serving as a coagulant chemical.

Sludge accumulated at the bottom of the sedimentation tank is a significant source of HAB cells and toxin if not managed correctly. For example, Pietsch found *Microcystis* sp. cell counts of 4,700,000 cells/L in clarifier sludge, significantly higher than the levels of 770,000 detected in the source water [9]. The most effective way to reduce the risk of cell resuspension or toxin release from sludge in the sedimentation basin is to increase the frequency of sludge withdrawals during active bloom periods. Laboratory experiments by Sun et al. suggest that microcystin may be released from *Microcystis* cells by natural cell lysis 6 days following alum coagulation and sedimentation [12] and perhaps as little as 2 days following coagulation by PACl [13]. Recycling sludge during active bloom periods should also be avoided. Consideration should be given regarding how the sludge is disposed of to prevent subsequent contamination of surface or groundwater.

6.3 Filtration

Filtration has long been used for drinking water treatment [16]. The process typically follows coagulation, flocculation, and sedimentation; however, direct filtration of surface water is also commonly practiced. The purpose of filtration is to remove fine particles or pathogens from drinking water that were not removed during coagulation, flocculation, and sedimentation. During filtration, water flows via gravity through a granular material, typically sand, anthracite, or garnet, and is collected in an underdrain system. Two types of filtration are generally practiced: rapid sand filtration and slow sand filtration. In rapid sand filtration, the filter units are periodically "backwashed" to maximize water flow through the filter. As a result, particle removal largely occurs via "depth filtration," which requires interception, impaction, or diffusion of particles for effective deposition rather than straining. Slow sand filtration, on the other hand, relies on the development of a biofilm layer, or "schmutzdecke," on the top of the filter to facilitate straining and the biological breakdown of filtered materials.

Research shows that HAB cells are only partially removed during rapid sand filtration. Furthermore, the cells that do accumulate on the filter media can be disrupted and act as a source of toxins to the finished water. For example, a study examining the removal of cyanobacteria in a full-scale drinking water treatment plant in southern Quebec found 16% to 26% breakthrough of cyanobacteria cells through a dual-media (sand and anthracite) rapid sand filter [17]. A variety of cyanobacteria were identified in the source water to this plant at the time of investigation, including *Microcystis* sp., *Anabaena* sp., *Pseudanabaena* sp., *Aphanothece* sp., *Aphanocapsa* sp., *Glaucospira* sp., *Planktolyngbya* sp., and *Aphanizomenon* sp., with *Microcystis* the most prevalent organism.

Because HAB cells accumulate in rapid sand filters, consideration should be given to backwashing conditions to minimize cell and toxin release. To reduce the potential for toxin release, filters should be backwashed more frequently during bloom periods. Backwash water disposal is also an issue to consider. Backwash water should not be recycled to the front of the treatment plant during HAB blooms. Surface or groundwater disposal of backwash water should be undertaken with caution to ensure no contamination of surface or groundwater.

HAB toxins are amendable to biodegradation; therefore, slow sand filtration is a potential viable treatment alternative if low hydraulic loading rates are acceptable. HAB cells are effectively removed during slow sand filtration, with removals of 95% for some toxins (e.g., microcystins) when an active schmutzdecke layer is present [18]. The operating conditions typically used for slow sand filters also raise potential challenges for ensuring low levels of toxins in the treated water. For example, the schmutzdecke layer accumulates HAB cells and therefore serves as a potential source for the release of toxins under conditions in which biodegradation is not active. Biodegradation may also occur in granular activated carbon (GAC) filters if a suitable biofilm is present.

Some research suggests that slow sand filters or GAC filters with an active biological layer might facilitate the biodegradation of HAB toxins. Ho et al. [19, 20], for example, examined the degradation of microcystin-LR and -LA in a biological sand filter and found no detectable toxins in the effluent following a 4-day lag period. Molecular analysis of the biofilm layer indicated the presence of the microcystin degradation gene cluster, specifically the mlA gene. Other researchers have also shown effective biodegradation of microcystins in slow sand filters, although degradation rates decreased at low temperature (<4°C) [18, 19]. For microcystins, the biodegradation products have been shown to be nontoxic [19]. The same is not true for saxitoxins. In one study at least, the biodegradation products of saxitoxin in a biologically activated filter were shown to be at least, or perhaps more, toxic than the parent compound [21].

6.4 Adsorption

Adsorption is a common process practiced in water treatment for the removal of micropollutants such as taste and odor compounds, herbicides and pesticides, and other various synthetic organic chemicals [3]. Given the effectiveness of adsorption for a broad range of contaminants, as well as the familiarity of the technology for water treatment, numerous adsorbents

FIGURE 6.3
Lake Erie harmful algal bloom and city of Toledo.

have been investigated in terms of effectiveness for removing HAB toxins. Adsorbents examined for their potential in removing HAB toxins include powdered activated carbon (PAC), GAC, carbon nanotubes [22], activated carbon fibers, iron oxide nanoparticles [23], and clay particles [24, 25].

The city of Toledo, Ohio, has used activated carbon for many years to treat Lake Erie water for taste and odors. More recently, however, the city of Toledo has turned to PAC to treat cyanotoxins, but at great expense. As shown in Figure 6.3, large HABs have occurred in Lake Erie near the city of Toledo. During the summer of 2012, it was estimated that the city of Toledo spent almost $200,000 per month to combat the combined *Microcystis* and diatom bloom. During normal operations to treat for taste and odor compounds, the city of Toledo spends about $1400 per day for PAC. During a bloom of *Microcystis*, however, the rate of carbon usage is about three times that needed for only taste and odor control. As a result, the city of Toledo spends anywhere from $6700 to $7500 per day for PAC. The net cost to treat for HAB toxins, therefore, is from $5300 to $6100 per day (or roughly $160,000 to $180,000 per month). The city of Toledo is not alone in incurring greater treatment costs when faced with HABs and cyanotoxins. The public water system in Clermont County, Ohio, estimated it spent approximately $600,000 to replace GAC to better deal with HABs.

PAC and GAC are the most commonly used sorbents in water treatment plants and find application for removing taste and odor compounds, disinfection by-product precursors, herbicides and pesticides, and other micropollutants [26]. Activated carbon is produced from a variety of raw materials (wood, coconut, other carbon sources) through a process of carbonation followed by slow oxidation. The resulting material has a highly porous structure consisting of micropores, mesopores, and macropores, which results in high specific surface areas. Activated carbon generally passing through an 80-mesh sieve is characterized as *powdered* activated carbon, while activated carbon retained on a 50-mesh sieve is termed *granular* activated carbon. PAC is typically dosed into the treatment stream of a water treatment plant to maximize contact time and is subsequently removed during liquid-solid separation processes such as sedimentation. GAC is utilized in packed columns.

The potential for using PAC and GAC to remove HAB toxins from drinking water has been reviewed a number of times (Svrcek [27], Hitzfeld [1], and Westrick [2]). Studies demonstrated the mesopore volume is an important factor controlling the extent of adsorption of HAB toxins to different types of activated carbon. Donati et al. [28], for example, showed that the adsorption density of microcystin-LR on eight different activated carbons correlated to the amount of mesopore volume and was unrelated to the micropore volume. The significance of mesopore volume can be understood by comparing the size of microcystin-LR relative to the diameter of mesopores. Mesopores are characterized as pores with diameters in the range of 2 to 50 nm; micropores are pores less than 2 nm. The size of microcystin-LR has been reported to be from 1.2 to 2.6 nm [28]. Therefore, microcystin-LR molecules may access adsorption sites within mesopores in the activated carbon but not micropores as they are generally too narrow.

Activated carbon produced from coconut shells has been found to have the greatest amount of micropore volume; wood- and coal-based carbons have greater mesopore and macropore volume [3]. The distribution of pore size in active carbon is typically attributed to the inherent physical structure of the raw material [3], with coconut shells possessing properties that result in mostly micropore volume during production. Although micropore volume is desirable for removal of smaller micropollutants, the relatively large size of HAB toxins like microcystins makes microporous carbon less effective. As a result, greater adsorption of microcystin-LR is observed with wood-based carbon compared to coconut-based materials [28, 29]. Owing to the smaller size of saxitoxins, they are more effectively removed by activated carbon having greater micropore volume [30].

Other measures of the characteristics of activated carbon, such as iodine number, Brunauer–Emmet–Teller (BET) surface area, and phenol number, generally do not correlate in a meaningful way to HAB toxin removal [27]. One study has shown that the point-of-zero charge of activated carbon influences adsorption capacity for HAB toxins [31]. Activated carbon with a higher point-of-zero charge maintains a more neutral or positive surface charge over a greater pH range, which leads to greater adsorption of microcystin-LR, which is negatively charged over the pH range observed in most natural waters.

HAB toxins and algal-derived taste and odor compounds, such as geosmin and methylisoborneol, may occur simultaneously in sources of drinking water. Given their varying adsorption behavior, the optimal type of carbon may not be the same for these different classes of contaminants. For micro-cystins, activated carbon with significant mesopore volume is needed; for small molecules like geosmin and methylisoborneol, activated carbon with micropore volume is sufficient. In such cases, water treatment plants will need to add multiple types of carbon for effective treatment of both classes of contaminants. For smaller HAB toxins like saxitoxins, a single microporous activated carbon may be sufficiently effective.

Microcystins are cyclic heptapeptides comprised of five fixed amino acids and two variable amino acids. Some of the common amino acids that make up the two units are leucine (L), arginine (R), tyrosine (Y), tryptophan (W), phenylalanine (F). Microcystin-LR, for example, is composed of the five fixed amino acids and leucine and tyrosine; microcystin-RR contains two arginine amino acids. The identity of the variable amino acids comprising a particular variant of microcystin influences the rate and extent of removal by activated carbon. For example, Cook and Newcombe showed that microcystin-RR was most effectively removed by activated carbon followed by microcystin-YR, -LR, and -LA [32, 33]. Ho et al. also showed that microcystin-LR was the most completely removed by activated carbon, followed by -YR, -LR, and -LA [34]. Pavagadhi et al. confirmed that microcystin-RR was adsorbed more quickly and to a greater extent on GAC and graphene oxide compared to microcystin-LR, at least at pH values greater than 5 [35]. Yan et al. examined the adsorption of microcystin-LR and -RR on a variety of adsorbents, includ-ing carbon nanotubes, activated carbon, talc, kaolinite, and sepiolite, and found -RR was again adsorbed to a greater extent than microcystin-LR [22]. The differences in adsorption of different microcystin variants are generally attributed to the differences in hydrophobicity, charge, and molecular size of the various microcystins.

Natural organic matter has a significant effect on the removal of HAB tox-ins by activated carbon. In general, organic matter reduces adsorption of micropollutants on activated carbon by direct competition for adsorption sites or by blocking access to pores. In studies with relatively small micro-pollutants, such as taste and odor compounds, pesticides, and herbicides, competition for adsorption sites is the dominant mechanism controlling adsorption [36]. As a result, the lower molecular weight fraction of natural organic matter is particularly important for the adsorption of small molec-ular weight micropollutants in micropores. Newcombe et al., for example, showed that the size of natural organic matter played an important role in controlling the extent of competition between taste and odor compounds (e.g., methylisoborneol) and natural organic matter for adsorption sites on activated carbon [37, 38]. They showed that the fraction of natural organic matter of similar size to the taste and odor compounds reduced adsorption most significantly. In their study, they showed that the adsorption of methyl-isoborneol was reduced most significantly by the ultrafiltration fraction of natural organic matter with a molecular weight less than 500.

A number of studies have documented how organic matter in source water reduces the rate and extent of HAB toxin removal by activated carbon [28, 31, 33, 34]. Donati et al. observed that the presence of natural organic matter significantly reduced the initial rate of microcystin-LR adsorption as well as the extent of adsorption [28]. In this early study, it was shown that the impact of natural organic matter on adsorption depended on the type of activated carbon. Huang et al. also found that the presence of natural organic matter reduced the adsorption capacity of activated carbon [31]. The percentage of microcystin-LR adsorbed decreased from 65% to 12% in the presence of natural organic matter. Using tannic acid as a model for natural organic matter, Campinas showed that, unlike smaller micropollutants, both competition and pore blockage are important mechanisms affecting the removal of microcystins by activated carbon in the presence of natural organic matter [39].

Less research has been conducted examining the removal of other HAB toxins, such as cylindrospermopsin, anatoxin-a, and saxitoxin, by activated carbon. It has been shown that cylindrospermopsin can be effectively removed by activated carbon. Ho et al. [34], for example, examined the removal of cylindrospermopsin by activated carbon and found the kinetics and extent of removal were similar to microcystin-RR. Newcombe and Nicholson [40] showed activated carbon is also effective at removing saxitoxin.

6.4.1 Modeling Powdered Activated Carbon Adsorption

A variety of techniques and approaches, varying in level of complexity, is available to estimate the dosage of activated carbon required to remove organic compounds, such as HAB toxins. The most widely used techniques or models are the Freundlich isotherm model, Langmuir model, the homogeneous surface diffusion model (HSDM) [33], ideal adsorbed solution theory (IAST), and the rapid column small-scale testing (RCSST) protocol.

The Freundlich model is an empirical approach assuming adsorption equilibrium between the sorbate (compound attaching to the surface, which in our case would be a HAB toxin) and the sorbent (or adsorbing surface). In the Freundlich model, the adsorption density (e.g., mass sorbate/mass sorbent) is related to the final or equilibrium concentration in solution by

$$q = K_f C_f^{1/n}$$

where q is the adsorption density, C_f is the final or equilibrium concentrate of contaminant in solution, and K_f and $1/n$ are empirically derived constants. Values for the Freundlich coefficients have been determined for a variety of contaminants and conditions (for a summary, see [26]). Combined with mass balance, the Freundlich model can be used to estimate the amount of activated carbon required to meet a desired treatment goal. To calculate the required dosage of PAC, the mass loss of contaminant (e.g., HAB toxin) from solution is assumed equal to the mass gain on the surface of the PAC. Therefore,

$$(C_0 - C)V = (q - q_0)m$$

where C_0 is the initial solution concentration of contaminant, C is the concentration of contaminant assumed to be in equilibrium with the sorbent after treatment with PAC, V is the volume of solution, and q_0 is the initial, sorbed amount of contaminant (typically assumed to be zero). If the initial concentration of contaminant on the sorbent is zero, then $q_0 = 0$. Combining the mass balance and Freundlich equations and solving for the dosage (m/V) of sorbent gives

$$\frac{m}{V} = \frac{C_0 - C}{K_f C^{1/n}}$$

which provides a straightforward approach for calculating the dosage of PAC needed to achieve a desired level of treatment and initial HAB toxin concentration.

A simple example illustrates the estimation of PAC dosages using the Freundlich approach and the relative effectiveness of wood- and coconut-based PAC. Consider, for example, an initial microcystin-LR concentration of 50 µg/L and a desired final concentration of 1 µg/L (consistent with the World Health Organization Provisional Guideline). Assume Freundlich isotherm constants for a wood-based PAC of K_f equal to 6309 $(\mu g/g)(L/\mu g)^{1/n}$ and $1/n$ equal to 0.56. For a coconut-based carbon, assume K_f of 1259 and $1/n$ of 1.0. The adsorption density for wood-based PAC is therefore

$$q = K_f C_f^{1/n}$$

$$q = (6309)(1)^{0.56}$$

$$q = 6309 \ \mu g/g$$

Based on the calculated adsorption density, the PAC dose needed to achieve a final microcystin-LR concentration of 1 µg/L is

$$\text{Dose} = \frac{C_0 - C}{q}$$

$$\text{Dose} = \frac{50-1}{6309} \times 1000 = 7.7 \text{ mg/L}$$

Similar calculations for coconut-based carbon give

$$q = (1259)(1)^{0.561}$$

$$q = 1259 \ \mu g/g$$

$$\text{Dose} = \frac{50-1}{1259} \times 1000 = 39 \text{ mg/L}$$

These calculations, based on published Freundlich isotherm constants, show that roughly five times more PAC is needed for the effective removal of microcystin-LR using coconut-based carbon compared to wood-based material.

With an estimate of the dosage needed to treat for cyanotoxins, the cost of carbon can be estimated for a hypothetical water treatment plant. Assuming a medium-size plant treating 50 million gallons per day (mgd) and a cost of activated carbon of $1700 per ton (current price), the cost of carbon per day can be estimated as

$$\text{Cost} = 7.7 \frac{\text{mg}}{\text{L}} \times 50 \times 10^6 \frac{\text{gal}}{\text{day}} \times \frac{\text{kg}}{10^6 \text{ mg}} \times 3.785 \frac{1}{\text{gal}} \times \frac{\text{ton}}{907 \text{ kg}} \times \$1700 \frac{1}{\text{ton}}$$

$$= \$2700/\text{day}$$

Therefore, assuming microcystin is present in the source water for 3 months during the summer, the total cost of carbon to treat for microcystin would be approximately $250,000. If a less-efficient coconut-based carbon (assuming dosage of 39 mg/L), the total cost of carbon for the summer would be approximately $1.2 million.

A major limitation of the Freundlich approach described is that it assumes a single solute in solution. As a result, the basic Freundlich approach does not take into account the role of competing solutes unless, of course, Freundlich isotherm parameters are determined in representative source waters.

Another approach to predicting multicomponent adsorption is the so-called IAST, originally adapted to aqueous solutions by Radke [41]. Few applications of IAST for HAB toxins have been conducted; however, additional work using this approach for predicting taste and odor compounds [38, 42, 43] and herbicides [43] has been carried out. In the IAST approach, an independent, single-solution equation describing the adsorption of each micropollutant in solution is developed based on the Freundlich isotherm equation. A set of n equations is then developed representing each micropollutant in solution. In environmental applications, a typical goal is to predict the adsorption of a particular micropollutant (e.g., geosmin or atrazine) in the presence of background organic matter, sometimes termed *equivalent background compound* (EBC).

Another limitation in the application of the Freundlich equation as described, including the extension using the IAST, is that the method assumes equilibrium sorption. In many cases, the contact time between activated carbon and HAB toxins may be short; therefore, it may not be correct to assume equilibrium conditions exist. In such cases, the kinetically limited adsorption density will be considerably less than that predicted for equilibrium conditions, and higher carbon dosages would be needed. A common approach for predicting the kinetics of adsorption of micropollutants on activated carbon is to use the HSDM. The HSDM assumes that the rate of adsorption of a pollutant on a sorbent is controlled by the rate of surface diffusion within the pores of the activated carbon. The model also assumes the solute is in local equilibrium between the sorbent and the interstitial fluid. The adsorption equilibrium can be determined using a variety of models, such as the Freundlich model. The mass transfer coefficients, including the surface diffusion coefficient D_s and the film mass transfer coefficient K_f, can be determined by various theoretical and applied approaches [26] or through lab- or pilot-scale testing.

The HSDM has been used to predict the adsorption of microcystin on PAC in the presence of background organic matter, specifically a model tannic acid [39]. In this research, IAST was used to predict the adsorption equilibrium between microcystin, tannic acid, and the PAC surface; the HSDM was used to elucidate the kinetics of adsorption. The results showed good agreement with experimental data and suggested that both pore blockage by tannic acid and competition for specific adsorption sites were important processes controlling the rate and extent of adsorption.

6.4.2 Modeling Granular Activated Carbon Adsorption

A similar approach based on the Freundlich model can be used to calculate the breakthrough time for a GAC column. In an adsorption column, water flows through the packed bed of GAC, providing the removal of contaminants via adsorption. Typically, the carbon first becomes "exhausted" at the inlet of the column and subsequently moves down the column until the entire amount of sorbent is consumed. Once the adsorbent is no longer able to adsorb contaminant, "breakthrough" occurs in which the concentration of contaminant at the outlet quickly increases to the inlet concentration. The key design goal is to determine how long the carbon will last until breakthrough occurs and the activated carbon must be replaced. Assuming a mass balance like we did for the PAC calculations, the adsorption density can be related to the mass solute adsorbed at breakthrough and the mass of carbon as

$$q = \frac{\text{mass solute adsorbed at breakthrough}}{\text{mass carbon}}$$

or

$$q = \frac{Q(C_0 - C/2)\tau_b}{m}$$

where Q is the water flow rate through the column, C_0 is the concentration of contaminant at the inlet of the column, C is the outlet concentration, t_b is the time to breakthrough, and m is the mass of carbon in the column.

We can consider a similar example as was used to predict the dosage of PAC required to reduce the concentration of microcystin from 50 to 1 µg/L. In considering the use of GAC, the goal is to determine how long the column of carbon will last until exhaustion. As an example, consider a GAC column with a design flow rate of 150,000 gallons/day, a filter area of 5 ft², and a filter depth of 3.0 ft. Assuming the same Freundlich constants as used for the PAC example, the adsorption density can be calculated as

$$q = (6309)(50)^{0.56}$$

$$q = 5.64 \times 10^4 \text{ mg/g}$$

It should be noted that, in the calculation of adsorption density, the outlet concentration of microcystin was used rather than the inlet concentration. The outlet concentration was used because the column is used until breakthrough occurs, at which time the activated carbon in the column is in equilibrium with the inlet concentration of microcystin. This varies compared to the calculation of adsorption density for the use of PAC. In the case of PAC, activated carbon is added to a mixing tank, which requires the PAC to be in equilibrium with the outlet concentration of microcystin that corresponds to the treatment goal, not the inlet concentration. As a result, the adsorption density achievable in a GAC column is often much higher than that possible for the application of GAC.

In the application of the Freundlich model to GAC columns, a safety factor is typically applied such that the actual adsorption density is assumed to be only 50% of the theoretical capacity as determined by the Freundlich equation. Therefore, for our example, the actual adsorption capacity is calculated as

$$q = 0.5 \times 5.64 \times 10^4 \text{ mg/g}$$

$$q = 2.82 \times 10^4 \text{ mg/g}$$

The mass of carbon in the column can be determined based on the dimensions of the column and a typical carbon density,

$$m = \rho V$$

where ρ is the density of carbon, and V is the volume of carbon in the GAC column. A typical value for the density of GAC is 38 lb/ft³. Therefore, the mass of carbon in the column can be calculated as

$$m = \left(38 \frac{lb}{ft^3} \right)\left(5.0 \ ft^2 \times 3.0 \ ft \right)\left(454 \frac{g}{lb} \right) = 2.6 \times 10^5 \ g$$

Now, with calculated values for the adsorption density and mass of carbon in the column, the time to breakthrough can be determined as

$$\tau_b = \frac{2.82 \times 10^4 \frac{mg}{g} \times 2.6 \times 10^5 \ g}{150,000 \frac{gal}{day} \times 3.785 \frac{1}{gal} \times \left(50 \frac{\mu g}{1} - \frac{1}{2} \frac{\mu g}{1} \right)} = 261 \ days$$

As for PAC, the cost of carbon for a GAC treatment system can be estimated. Based on the calculations, 260 kg of GAC are needed every 261 days to treat 150,000 gallons per day. For comparison to PAC, assuming a flow rate of 50 mgd and 90 days of treatment during the summer, the total cost of GAC for a HAB season would be

$$\text{Cost} = 260 \ kg \times \frac{50 \ mgd}{0.15 \ mgd} \times \frac{90 \ days}{261 \ days} \times \frac{ton}{907 \ kg} \times \frac{\$1700}{ton} = \$56,000$$

We can do similar calculations for the coconut-based carbon, in which case the same column treating 150,000 gallons per day would be sufficient to treat water for 53 days because of the lower adsorption capacity. In this case, considering a larger treatment plant requiring 50 mgd, the cost of coconut-based carbon would be

$$\text{Cost} = 260 \ kg \times \frac{50 \ mgd}{0.15 \ mgd} \times \frac{90 \ days}{53 \ days} \times \frac{ton}{907 \ kg} \times \frac{\$1700}{ton} = \$275,000$$

Therefore, the cost for wood-based carbon alone in a GAC system to treat a microcystin bloom during a 3-month period is approximately $56,000, which is less than that required for PAC because of the greater achievable adsorption density in GAC systems. However, GAC systems require greater capital costs than PAC systems, so a water utility would need to examine the trade-offs in capital versus operational costs. If a GAC plant utilized the less-efficient coconut-based carbon, the total cost for the summer would be $275,000, which is comparable to the cost of PAC using the more efficient wood-based material. In all cases, however, the costs incurred for treating HAB toxins using PAC or GAC are substantial and likely to put a significant financial strain on any water utility.

The Freundlich isotherm approach, for both PAC and GAC, does not capture many of the factors that influence the performance of activated carbon systems at full scale. The calculations illustrated, for example, do not consider competing substances, such as taste and odor compounds or dissolved natural organic matter, both of which may reduce the effectiveness of PAC for removing microcystin. Also, Freundlich isotherm constants are typically determined at the laboratory scale and therefore do not capture some of the mass transfer and other limitations observed at larger scales and applications.

To better predict the performance of activated carbon systems, a number of scale-up methodologies have been developed [3]. One approach for predicting the performance of pilot- or full-scale GAC systems is the RSSCT procedure. The RSSCT procedure is based on the dispersed flow pore surface diffusion model [3]. Based on this model, a number of nondimensional numbers can be identified that are conserved going from the lab to full scale. Assuming constant diffusivity at different scales, for example, the following relation can be established:

$$\frac{EBCT_{sc}}{EBCT_{lc}} = \frac{d_{sc}^2}{d_{lc}^2} = \frac{\tau_{sc}}{\tau_{lc}}$$

where $EBCT$ is the empty bed contact time, and d is the sorbent (e.g., activated carbon) diameter. The subscripts sc and lc refer to the small column (i.e., lab-scale) and large column (i.e., full-scale) conditions, respectively.

The nondimensional analysis suggests a procedure for accurately predicting the breakthrough of a full-scale GAC column based on laboratory data. In the RSSCT procedure, the sorbent (e.g., GAC) is crushed to a fraction of the sorbent size expected to be used for the full-scale process. Based on the nondimensional analysis presented, a test column can be designed using the ratio of the square of the particle diameters and the calculated EBCT for the small column. Lab-scale column tests are then carried out to determine the time to breakthrough for the small-scale column. The time to breakthrough for the full-scale column can then be predicted based on the small-scale data.

Consider, for example, the application of the RSSCT method to predict the performance of a full-scale GAC column operating at an EBCT of 7.5 minutes, given the design parameters shown in Table 6.1. The GAC to be used in the full-scale column is chosen to have a diameter of 1.0 mm. After

TABLE 6.1

Model Parameters for RSSCT Procedure

Design Parameter	Unit	Large Scale	Small Scale
GAC particle diameter	Millimeters	1.0	0.21
EBCT	Minutes	7.5	0.33
Hydraulic loading rate	Meters/hour	5.0	23.8
Column length	Centimeters	62.5	13.1
Column diameter	Centimeters	—	1.0
Bed volumes to breakthrough		—	15,000
t_b	Days	78.1	3.4

crushing, the diameter of the GAC to be used in the small-scale column is 0.21 mm. Based on similitude, the EBCT for the small-scale column can be determined as

$$EBCT_{sc} = EBCT_{lc}\frac{d_{sc}^2}{d_{lc}^2}$$

$$EBCT_{sc} = 7.5 \times \frac{(0.21)^2}{(1.0)^2} = 0.33 \text{ min}$$

If the hydraulic loading rate of the full-scale column is chosen as 5.0 m/h, then by similitude the hydraulic loading rate of the small-scale column v_{sc} can be calculated as

$$v_{sc} = v_{lc}\frac{d_{lc}}{d_{sc}} = 5.0 \times \frac{1.0}{0.21} = 23.8 \text{ m/h}$$

With an estimate of the hydraulic loading rate for the small-scale column and EBCT, the length of the small-scale column can be calculated as

$$L_{sc} = v_{sc}EBCT_{sc} = \frac{(23.8 \text{ m/h})(1000 \text{ mm/m})(0.33 \text{ min})}{(60 \text{ min/h})} = 131 \text{ mm}$$

For comparison, the length of the full-scale column can also be calculated based on the full-scale hydraulic loading rate and EBCT and is determined to be 62.5 cm, or about 2 ft. The flow rate needed for the small-scale column test to achieve the required hydraulic loading rate can be calculated, assuming a column diameter of 1.0 cm, as

$$Q_{sc} = v_{sc}A_{sc} = \frac{(23.8 \text{ m/h})(3.14)(1.0 \text{ cm})^2(100 \text{ cm/m})}{(4´60 \text{ min/h})} \times \frac{\text{mL}}{\text{cm}^3} = 31.1 \text{ mL/min}$$

where A_{sc} is the cross-sectional area of the small-scale column. Now that the small-scale column has been "scaled down," the laboratory column test can be conducted using the actual water to be treated by the full-scale system. The small-scale column tests can be used to estimate the effectiveness of treatment as well as the duration of treatment until breakthrough occurs. As a hypothetical example, assume a small-scale column test was carried out and approximately 15,000 bed volumes (BVs) were treated before break-through of microcystin occurred. For this case, the time to breakthrough for the small-scale column would have occurred in

$$\tau_{sc} = BV \times EBCT_{sc} = 15,000 \times 0.33 \text{ min} = 4950 \text{ min}$$

or approximately 3.4 days. Based on similitude, the full-scale column would be expected to treat a similar BV before breakthrough; therefore, the expected time until breakthrough for the full-scale column would be

$$\tau_{lc} = BV \times EBCT_{lc} = 15,000 \times 7.5 \text{ min} = 112,500 \text{ min}$$

or approximately 78 days. The major benefit of the RSSCT is clear: The break-through of the full-scale column can be predicted using the small-scale laboratory test in the matter of a few days, rather than weeks or months. Other benefits of the RSSCT procedure are that it is based on a theoretical foundation and can be carried out utilizing the actual water to be treated. One potential limitation of the approach is that the crushing of the GAC may introduce nonscalable changes to the carbon and various transport pro-cesses [44]. Also, the short duration of the column tests make it difficult to simulate long-term changes in water quality and predict processes that may occur over longer durations, such as biofouling of the carbon and biodegra-dation of the contaminants.

At least one research article has examined the use of the RSSCT procedure for predicting the performance of GAC for removing HAB toxins [45]. Hall et al. examined the removal of microcystin and anatoxin-a using GAC and the RSSCT procedure. In the laboratory tests, the authors found that anatoxin-a was removed slightly better than microcystin using GAC. Using the RSSCT procedure, Hall et al. found that significant breakthrough of microcystin and anatoxin-a occurred (80% of influent concentration) after about 30,000 BVs, which represented a breakthrough time of approximately 18 weeks at the full scale. The authors speculated that the short time to breakthrough was a result of the constant influent concentration of microcystin and anatoxin-a used in the lab-scale tests and suggested that times to breakthrough for the full-scale systems would be longer if the columns were not continuously challenged with the toxins.

The short bed adsorption (SBA) test is another approach developed to pre-dict the breakthrough of HAB toxins using GAC [44]. The SBA test utilizes laboratory-scale column tests to obtain basic kinetic data that can be used to predict pilot- or full-scale performance based on the HSDM. The SBA test was first developed and applied for predicting the breakthrough of pesticides using GAC. The advantages of the SBA procedure are that it uses the same GAC as the pilot- or full-scale columns (without grinding), and perhaps more important, the GAC can be preloaded prior to use in the SBA test. Ho and Newcombe used the SBA procedure to predict the breakthrough of microcystin LA and microcystin LR on virgin and preloaded GAC utilizing source water from the Myponga Water Treatment Plant in Southern Australia [44]. The carbon used was a wood-based carbon (picazine) preloaded in the pilot-scale system for a period of 6 months.

To carry out the SBA procedure, Freundlich parameters were determined using the bottle-point technique, and kinetic coefficients were ascertained using a 1.0-cm diameter, 30 cm long glass column. The hydraulic loading rate between the lab-scale and pilot-scale systems was conserved at a rate of 9.9 cm/hour. Breakthrough curves from the lab-scale column were used to estimate the two kinetic parameters needed for the HSDM, namely, the surface diffusion coefficient D_s and the film mass transfer coefficient k_f. Column data were fitted using a nonlinear, least-squares optimization technique. The SBA test results and subsequent modeling using the HSDM predicted that 90% breakthrough of HAB toxins at the pilot-scale facility should occur after 6 months of operation. This prediction, however, did not correspond to pilot-scale data, which showed little or no breakthrough after 6 months. Like other studies, the authors suggested that the lack of breakthrough of HAB toxins was caused by biodegradation in the columns that effectively increased the lifespan of the GAC. Ho and Newcombe performed additional biodegradation experiments, which provided some evidence supporting this conclusion about the significance of biodegradation during GAC treatment of HAB toxins.

6.5 Chemical Disinfection

The disinfection process for drinking water treatment evolved over the late 1800s to early 1900s to eliminate disease-causing organisms. Chlorine (including free chlorine, combined chlorine, and chlorine dioxide) and ozone are the two most widely used disinfectants in drinking water treatment. Both chlorine and ozone are strong oxidants and are capable of not only inactivating microorganisms but also oxidizing micropollutants. When considering the application of disinfection technologies for the control of HAB and HAB toxins in drinking water, there are two important considerations: (1) What are the required dose and contact time needed to sufficiently destroy specific HAB organisms and toxins? and (2) To what extent will disinfection disrupt HAB cells and lead to the release of intercellular toxin into the treated water?

The effectiveness of chemical disinfection for destroying HAB toxins in drinking water depends on the particular toxin; for chlorine, the type of chlorine compound used is also important. Furthermore, the level of destruction for a particular toxin depends on the contact time, dosage, and source water quality. Free chlorine is generally effective at destroying microcystins and cylindrospermopsin but less effective in the treatment of anatoxin-a and saxitoxins. Chloramines (i.e., chlorine combined with ammonia) and chlorine dioxide are not effective for the destruction of HAB toxins. Ozone is effective at destroying microcystins, cylindrospermopsin, and anatoxin-a but not saxitoxins. As can be seen, there is no single chemical disinfectant effective for treatment of all main classes of HAB toxins.

The effect of chemical dosage and contact on the effectiveness of chemical disinfection is typically modeled using the Chick–Watson Law, where

$$C^n t = \text{constant}.$$

In this expression, the product of the concentration of disinfectant C and time t is equal to a constant for a given level of inactivation, and n is a constant. Thus, a desired level of inactivation can be achieved through the appropriate selection of dosage and contact time. Using a higher dosage allows for a shorter contact time and vice versa. In many cases at water treatment plants, the contact time is relatively fixed because of the dimensions of the clear well, flow rate, and time of travel in the distribution system. In such cases, the dosage of chemical disinfectant is selected to achieve the desired level of inactivation for a fixed contact time. For the disinfection of microorganisms, such as viruses, *Giardia*, and *Cryptosporidium*, regulatory agencies have tabulated values for different levels of inactivation (so-called "CT Tables") for various chemical disinfectants. The United States Environmental Protection Agency (USEPA), for example, has developed CT tables in support of the surface water treatment rule (SWTR). The CT tables typically provide required CT values to achieve a desired level of inactivation. The level of inactivation is represented in terms of "log removal." A 1-log removal represents 90.0% inactivation, 2-log removal represents 99.0% inactivation, 3-log removal represents 99.9% inactivation, and so on.

Understanding the CT concept for disinfection is useful when considering the application of chlorine for the destruction of HAB toxins. Therefore, it is helpful to consider the application of chlorine for the disinfection of *Giardia* as an example before discussing the more specific case of HAB toxins. For the inactivation of *Giardia* by chlorine, the CT values depend on the dosage of chlorine, pH, and level of disinfection. To achieve 3.0 log removal of *Giardia* at pH 6.0, for example, a free chlorine concentration of 0.8 mg/L requires a CT value of 145 mg/L-minute. To achieve 1-log removal at this same pH and free chlorine concentration requires only 48 mg/L-minute. The CT values reflect the dependence of chlorine inactivation on solution pH because

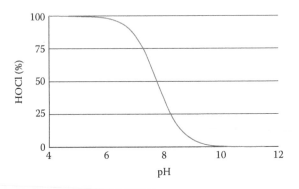

FIGURE 6.4
Distribution of chlorine species as a function of pH.

higher CT values are required to achieve a desired level of inactivation as the pH increases. Chlorine is less effective as a disinfectant as the pH increases because of the acid-base properties of hypochlorous acid. Hypochlorous acid is a weak acid and dissociates around pH values that are slightly higher than neutral, as shown in Figure 6.4. The dissociation is described by the following acid-base reaction:

$$HOCl = H^+ + OCl^-$$

HOCl as a disinfectant has many times greater strength compared to OCl⁻; therefore, disinfection is more effective using chlorine at lower pH values that favor the presence of HOCl.

The pH-dependent dissociation of HOCl can be evaluated based on the pK_a value for the reaction shown. The pK_a value for this reaction depends on temperature and can be determined based on the following equation:

$$pK_a = \frac{3000.00}{T} - 10.0686 + 0.0253T$$

where T is in degrees kelvin. To illustrate the use of this equation, the pK_a at a temperature of 0.5°C (278 K) is calculated as

$$pK_a = \frac{3000.00}{278} - 10.0686 + 0.0253 \times 278$$

$$pK_a = 7.76.$$

This value of the pK_a demonstrates that, at a temperature of 0.5°C, the concentrations of HOCl and OCl⁻ are equivalent at a pH value of 7.7. At lower pH values, HOCl is the dominant species; at pH values higher than 7.7, OCl⁻ is more prevalent.

Consider now the distribution of HOCl in Figure 6.4 and the USEPA CT values for *Giardia*. Previously, the CT value for 3-log removal of *Giardia* at a temperature of 0.5°C or lower was determined as 145 mg/L-minute. At this pH value, nearly all the chlorine is in the form of HOCl. Now, consider the CT value required for 3-log removal of *Giardia* at a temperature of 0.5°C or lower at a pH value of 9.0. At this pH value, nearly all the chlorine is in the form of the less-effective disinfectant OCl⁻. From the USEPA CT table, the required CT value at pH 9.0 is 422 mg/L-minute, nearly three times greater than the CT value needed at pH 6.0.

The CT concept is also valuable in assessing the effectiveness of various chemical disinfectants for the destruction of HAB toxins. Acero et al., for example, developed CT values for the elimination of microcystins by chlorine [46] based on the pseudo-first-order rate constant data (see Figure 6.5). CT values for the elimination of microcystin-LR by chlorine in a completely stirred tank reactor (CSTR) are reproduced from Acero et al. in Table 6.2. Like CT values for the inactivation of *Giardia* and other microorganisms, the CT values for the elimination of microcystin-LR depend on the temperature

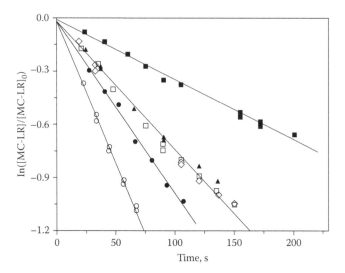

FIGURE 6.5
Pseudo-first-order kinetic plot for the chlorination of microcystin-LR (MC-LR) at 20°C and pH 7.2. Initial concentrations: $[HOCl]_0 = 49.2\ \mu M$, $[MC\text{-}LR]_0 = 2.35\ \mu M$ (■); $[HOCl]_0 = 95.7\ \mu M$, $[MC\text{-}LR]_0 = 2.30\ \mu M$ (□); $[HOCl]_0 = 142.5\ \mu M$, $[MC\text{-}LR]_0 = 2.35\ \mu M$ (●); $[HOCl]_0 = 222.8\ \mu M$, $[MC\text{-}LR]_0 = 2.50\ \mu M$ (○); $[HOCl]_0 = 96.2\ \mu M$, $[MC\text{-}LR]_0 = 1.49\ \mu M$ (▲); $[HOCl]_0 = 94.6\ \mu M$, $[MC\text{-}LR]_0 = 1.91\ \mu M$ (◊). (From J.A. Westrick, D.C. Szlag, B.J. Southwell, and J. Sinclair, *Analytical and Bioanalytical Chemistry*, 397 (2010) 1705–1714.)

TABLE 6.2

CT Values for Oxidation of Microcystin-LR to 1 µg/L Using Free Chlorine

pH	Initial Microcystin-LR (µg/L)	CT Value (mg L⁻¹ minute)			
		10°C	15°C	20°C	25°C
6	50	583.9	503.3	436.3	380
	10	107.2	92.4	80.1	69.8
7	50	847.7	731.2	663.7	551.7
	10	155.7	134.3	116.4	101.3
8	50	2347.5	2020.3	1751.8	1525.9
	10	431.2	371.1	321.7	280.3
9	50	7731.1	6589	5740.9	4998.6
	10	1420	1210.2	1054.4	918.1

Source: Reproduced from J.L. Acero, E. Rodriguez, and J. Meriluoto, *Water Research*, 39 (2005) 1628–1638.

and pH, with lower CT values required at lower pH and higher temperature. The CT values shown in Table 6.2 present the product of C and T required to reduce the microcystin-LR concentration to 1 µg/L from a starting microcystin-LR concentration of either 50 or 10 µg/L. The required CT value to reduce the concentration of microcystin-LR from 50 to 1 µg/L at pH 6 and 25°C, for example, is 380.0 mg/L-minute. At pH 9 and 25°C, the required CT value increases to 4998.6 mg/L-minute.

As an example, consider the contact time needed to reduce microcystin-LR from 50 to 1 µg/L at a pH of 6 and temperature of 25°C. Assuming a free chlorine concentration of 1.5 mg/L, the required contact time is

$$T = \frac{CT}{C}$$

$$T = \frac{380 \frac{mg}{L} - min}{1.5 \ mg/L}$$

$$T = 253 \ minutes$$

At pH 9, the concentration of HOCl is lower and the CT value required is significantly higher. As a result, the calculated contact time T increases to 3332.4 minutes. These calculations demonstrate that chemical disinfection with chlorine may be effective at destroying toxins at low pH values (significantly below the pK$_a$), but relatively long contact times are needed. At high pH values, chlorine is not an effective approach for destroying toxins.

6.6 Other Disinfection Technologies

Both ultraviolet (UV) disinfection and ozone or common technologies used for the disinfection of drinking water and significant research have explored the use of these technologies for HABs and HAB toxins. UV disinfection, ozone, and "advanced oxidation technologies" are the focus of the next chapter on advanced treatment approaches.

References

1. B.C. Hitzfeld, S.J. Hoger, and D.R. Dietrich, Cyanobacterial toxins: removal during drinking water treatment, and human risk assessment. *Environmental Health Perspectives*, 108 (2000) 113–122.
2. J.A. Westrick, D.C. Szlag, B.J. Southwell, and J. Sinclair, A review of cyanobacteria and cyanotoxins removal/inactivation in drinking water treatment. *Analytical and Bioanalytical Chemistry*, 397 (2010) 1705–1714.
3. J.C. Crittenden, R.R. Trussell, D.W. Hand, K.J. Howe, and G. Tchobanoglous, *Water Treatment: Principles and Design*, 2nd ed. Hoboken, NJ: Wiley, 2005.
4. R.K. Henderson, S.A. Parsons, and B. Jefferson, Successful removal of algae through the control of zeta potential. *Separation Science and Technology*, 43 (2008) 1653–1666.
5. B. Kwon, N. Park, and J. Cho, Effect of algae on fouling and efficiency of UF membranes. *Desalination*, 179 (2005) 203–214.
6. R. Henderson, S.A. Parsons, and B. Jefferson, The impact of algal properties and pre-oxidation on solid–liquid separation of algae. *Water Research*, 42 (2008) 1827–1845.
7. S. Hadjoudja, V. Deluchat, and M. Baudu, Cell surface characterisation of *Microcystis aeruginosa* and *Chlorella vulgaris. Journal of Colloid and Interface Science*, 342 (2010) 293–299.
8. M.R. Teixeira and M.J. Rosa, Comparing dissolved air flotation and conventional sedimentation to remove cyanobacterial cells of *Microcystis aeruginosa*: Part I: The key operating conditions. *Separation and Purification Technology*, 52 (2006) 84–94.
9. J. Pietsch, K. Bornmann, and W. Schmidt, Relevance of intra- and extracellular cyanotoxins for drinking water treatment. *Acta hydrochimica et hydrobiologica*, 30 (2002) 7–15.
10. J. Rositano and B.C. Nicholson, *Water Treatment Techniques for Removal of Cyanobacterial Toxins from Water*. Australian Centre for Water Quality Research, Salisbury, South Australia, 1994.
11. C.W.K. Chow, M. Drikas, J. House, M.D. Burch, and R.M.A. Velzeboer, The impact of conventional water treatment processes on cells of the cyanobacterium *Microcystis aeruginosa. Water Research*, 33 (1999) 3253–3262.

12. F. Sun, H.-Y. Pei, W.-R. Hu, and C.-X. Ma, The lysis of *Microcystis aeruginosa* in AlCl3 coagulation and sedimentation processes. *Chemical Engineering Journal*, 193–194 (2012) 196–202.

13. F. Sun, H.-Y. Pei, W.-R. Hu, X.-Q. Li, C.-X. Ma, and R.-T. Pei, The cell damage of *Microcystis aeruginosa* in PACl coagulation and floc storage processes. *Separation and Purification Technology*, 115 (2013) 123–128.

14. H.G. Peterson, S.E. Hrudey, I.A. Cantin, T.R. Perley, and S.L. Kenefick, Physiological toxicity, cell membrane damage and the release of dissolved organic carbon and geosmin by *Aphanizomenon flos-aquae* after exposure to water treatment chemicals. *Water Research*, 29 (1995) 1515–1523.

15. B.-L. Yuan, J.-H. Qu, and M.-L. Fu, Removal of cyanobacterial microcystin-LR by ferrate oxidation–coagulation. *Toxicon*, 40 (2002) 1129–1134.

16. M.N. Baker, *The Quest for Pure Water: The History of Water Purification from the Earliest Records to the Twentieth Century*. American Water Works Association, New York, 1949.

17. A. Zamyadi, S.L. MacLeod, Y. Fan, N. McQuaid, S. Dorner, S. Sauvé, and M. Prévost, Toxic cyanobacterial breakthrough and accumulation in a drinking water plant: a monitoring and treatment challenge, *Water Research*, 46 (2012) 1511–1523.

18. G. Grützmacher, G. Böttcher, I. Chorus, and H. Bartel, Removal of microcystins by slow sand filtration. *Environmental Toxicology*, 17 (2002) 386–394.

19. L. Ho, E. Sawade, and G. Newcombe, Biological treatment options for cyanobacteria metabolite removal—a review. *Water Research*, 46 (2012) 1536–1548.

20. L. Ho, T. Meyn, A. Keegan, D. Hoefel, J. Brookes, C.P. Saint, and G. Newcombe, Bacterial degradation of microcystin toxins within a biologically active sand filter. *Water Research*, 40 (2006) 768–774.

21. N. Kayal, G. Newcombe, and L. Ho, Investigating the fate of saxitoxins in biologically active water treatment plant filters. *Environmental Toxicology*, 23 (2008) 751–755.

22. H. Yan, A. Gong, H. He, J. Zhou, Y. Wei, and L. Lv, Adsorption of microcystins by carbon nanotubes. *Chemosphere*, 62 (2006) 142–148.

23. J. Lee and H.W. Walker, Adsorption of microcystin-Lr onto iron oxide nanoparticles. *Colloids and Surfaces A–Physicochemical and Engineering Aspects*, 373 (2011) 94–100.

24. R.J. Morris, D.E. Williams, H.A. Luu, C.F.B. Holmes, R.J. Andersen, and S.E. Calvert, The adsorption of microcystin-LR by natural clay particles. *Toxicon*, 38 (2000) 303–308.

25. J.M. Burns, S. Hall, and J.L. Ferry, The adsorption of saxitoxin to clays and sediments in fresh and saline waters. *Water Research*, 43 (2009) 1899–1904.

26. R. Sykes and H.W. Walker, Physical water and wastewater treatment processes, in: W.F. Chen and J.Y.R. Liew (Eds.), *The Civil Engineering Handbook*. New York: CRC Press, 2003, pp. 9-1 to 9-141.

27. C. Svrcek and D.W. Smith, Cyanobacteria toxins and the current state of knowledge on water treatment options: a review. *Journal of Environmental Engineering and Science*, 3 (2004) 155–185.

28. C. Donati, M. Drikas, R. Hayes, and G. Newcombe, Microcystin-LR adsorption by powdered activated carbon. *Water Research*, 28 (1994) 1735–1742.

29. J. Lee and H.W. Walker, Effect of process variables and natural organic matter on removal of microcystin-LR by PAC-UF. *Environmental Science and Technology*, 40 (2006) 7336–7342.

30. G. Newcombe, *Removal of Algal Toxins from Drinking Water Using Ozone and GAC*. American Water Works Association Research Foundation, Denver, CO, 2002.

31. W.-J. Huang, B.-L. Cheng, and Y.-L. Cheng, Adsorption of microcystin-LR by three types of activated carbon. *Journal of Hazardous Materials*, 141 (2007) 115–122.

32. D. Cook and G. Newcombe, Removal of microcystin variants with powdered activated carbon. *Water Supply*, 2 (2002) 201–207.

33. D. Cook and G. Newcombe, Comparison and modeling of the adsorption of two microcystin analogues onto powdered activated carbon. *Environmental Technology*, 29 (2008) 525–534.

34. L. Ho, P. Lambling, H. Bustamante, P. Duker, and G. Newcombe, Application of powdered activated carbon for the adsorption of cylindrospermopsin and microcystin toxins from drinking water supplies, *Water Research*, 45 (2011) 2954–2964.

35. S. Pavagadhi, A.L.L. Tang, M. Sathishkumar, K.P. Loh, and R. Balasubramanian, Removal of microcystin-LR and microcystin-RR by graphene oxide: adsorption and kinetic experiments. *Water Research*, 47 (2013) 4621–4629.

36. J.E. Kilduff, T. Karanfil, and W.J. Weber, Jr., Competitive effects of nondisplaceable organic compounds on trichloroethylene uptake by activated carbon. II. Model verification and applicability to natural organic matter. *Journal of Colloid and Interface Science*, 205 (1998) 280–289.

37. G. Newcombe, M. Drikas, and R. Hayes, Influence of characterised natural organic material on activated carbon adsorption: II. Effect on pore volume distribution and adsorption of 2-methylisoborneol. *Water Research*, 31 (1997) 1065–1073.

38. G. Newcombe, J. Morrison, C. Hepplewhite, and D.R.U. Knappe, Simultaneous adsorption of MIB and NOM onto activated carbon: II. Competitive effects. *Carbon*, 40 (2002) 2147–2156.

39. M. Campinas, R.M.C. Viegas, and M.J. Rosa, Modelling and understanding the competitive adsorption of microcystins and tannic acid. *Water Research*, 47 (2013) 5690–5699.

40. G. Newcombe and B. Nicholson, Treatment options for the saxitoxin class of cyanotoxins. *Water Supply*, 2 (2002) 271–275.

41. C.J. Radke and J.M. Prausnitz, Thermodynamics of multi-solute adsorption from dilute liquid solutions. *AIChE Journal*, 18 (1972) 761–768.

42. M.R. Graham, R.S. Summers, M.R. Simpson, and B.W. MacLeod, Modeling equilibrium adsorption of 2-methylisoborneol and geosmin in natural waters. *Water Research*, 34 (2000) 2291–2300.

43. D.R.U. Knappe, Y. Matsui, V.L. Snoeyink, P. Roche, M.J. Prados, and M.-M. Bourbigot, Predicting the capacity of powdered activated carbon for trace organic compounds in natural waters. *Environmental Science and Technology*, 32 (1998) 1694–1698.

44. L. Ho and G. Newcombe, Evaluating the adsorption of microcystin toxins using granular activated carbon (GAC). *Journal of Water Supply: Research and Technology–AQUA*, 56 (2007) 281–291.

45. T. Hall, J. Hart, B. Croll, and R. Gregory, Laboratory-scale investigations of algal toxin removal by water treatment. *Water and Environment Journal*, 14 (2000) 143–149.

46. J.L. Acero, E. Rodriguez, and J. Meriluoto, Kinetics of reactions between chlorine and the cyanobacterial toxins microcystins. *Water Research*, 39 (2005) 1628–1638.

7

Advanced Treatment Processes for the Removal of HAB Cells and Toxins from Drinking Water

7.1 Introduction

Although conventional treatment processes can be effective for the removal for harmful algal bloom (HAB) cells, and for some HAB toxins under optimal conditions, the potential exists for significant breakthrough of toxins during normal operation. As a result, there is a recognized need for more advanced treatment processes for the treatment of HAB toxins in drinking water. Membrane technologies are increasingly being used for the purification of drinking water, from both freshwater sources and the ocean. As a result, there is interest in understanding the capabilities of these processes for the removal of HAB cells and HAB toxins. The use of membrane processes, alone or in combination with other technology, such as coagulation and powdered activated carbon (PAC), is of increasing interest. Along with more traditional oxidation approaches (e.g., ozone and potassium permanganate), advanced oxidation processes (AOPs) are a promising suite of technologies that offer advantages for the destruction of HAB toxins in drinking water. A number of AOPs are currently under development and examined for destroying HAB toxins, including photolysis, ultraviolet (UV)/H_2O_2, UV/TiO_2, technologies based on generation of sulfate radical, ultrasound, Fenton reaction, and ferrate. This chapter provides an introduction to membrane and oxidation technologies and their application to removing or destroying HAB toxins in drinking water.

7.2 Membrane Processes

Novel membrane processes, including reverse osmosis (RO), nanofiltration (NF), ultrafiltration (UF), and microfiltration (MF) are increasingly being developed and applied to meet more stringent drinking water quality requirements and the need for the treatment of multiple contaminants of concern. A typical membrane treatment process consists of a membrane unit, feed pump, pretreatment system, posttreatment system, and backwash system. The pretreatment system may consist of preliminary separation processes or pH adjustment to prevent or minimize fouling in the membrane module. The posttreatment system may be needed for disinfection and to adjust pH or other water quality parameters to minimize corrosion or scaling in the distribution system. Periodically, membrane systems must be backwashed to reduce fouling; therefore, some type of backwashing system is generally part of the total membrane treatment package. The feed pump provides the necessary driving force for transport of water through the membrane pores.

Typical characteristics of various membrane processes are given in Table 7.1. The nominal pore size of membranes varies widely, with RO systems essentially nonporous and MF membranes having a pore size on the order of 0.1 μm. RO membranes are capable of rejecting most molecules and allowing only the passage of water through the material. NF membranes have typical molecular weight cutoffs (MWCOs) of approximately 1000 Da. MF membranes have pore sizes with MWCOs on the order of 50,000 or greater. The operating pressure generally increases with decreasing pore size, with RO membranes requiring an operating pressure of greater than 1000 psi in some cases. NF membranes operate in the range of 100–500 psi, and UF and MF require operating pressures of 40 psi or less. It should be noted that the characteristic values given in Table 7.1 represent typical values and size; MWCO and operating pressure can vary significantly for different membrane types.

Membrane modules come in a number of different configurations, including flat sheet, hollow fibers, spiral round, and others. The two most common configurations for water treatment applications are spiral wound and hollow-fiber modules. The modular nature of membrane units is one of the major advantages of this technology because it allows for relatively easy expansions in capacity and system updating. A variety of materials is used to manufacture membranes, including cellulose acetate, various modified cellulose acetates, polysulfone, polyethersulfone, polyamide, polyvinylidene fluoride, and others. Each membrane material possesses different physicochemical properties, such as hydrophobicity/hydrophilicity, zeta potential,

TABLE 7.1

Characteristics of Various Membrane Processes

Membrane Technology	Nominal Pore Size (μm)	Molecular Weight Cutoff	Typical Operating Pressures (psi)
Reverse osmosis	Nonporous < 0.0005	<500	500–1200
Nanofiltration	0.001	100–500	100–500
Ultrafiltration	0.01	50,000	20–40
Microfiltration	0.1	>50,000	3–20

and chemical resistance, to name a few. The differences in properties have a significant impact on the performance of membrane systems for drinking water applications. Thin-film composite membranes are common; the membrane consists of a thin outer layer in contact with the feed and a more porous backing layer to improve mass transfer.

A number of different factors and processes control the rejection of contaminants in membrane systems (for a review, see [1]). The rejection of organic compounds in membrane systems may be caused by size exclusion (sometimes referred to as steric hindrance), nonspecific interactions with the membrane surface, effects related to membrane fouling, and other processes. When the size of the pores is smaller than the solute, steric hindrance is typically considered to play a major role in controlling contaminant rejection. However, other nonspecific interactions, such as electrostatic or hydrophobic interactions, may also play a role in controlling contaminant rejection in membrane systems. For example, favorable electrostatic attraction between contaminants and membrane surfaces may result in contaminant adsorption on the membrane surface. Similarly, hydrophobic interactions may also promote adsorption and reduce the concentration of contaminant in the permeant, at least at early operational times. Other factors, such as concentration-polarization, may also influence the flux of contaminants, such as HAB toxins, in membrane systems.

7.2.1 RO and NF for Removal of HABs and HAB Toxins

Reverse osmosis and NF can effectively remove HAB cells and, in many cases, toxins as well. As shown in Table 7.1, for example, the typical pore size of NF membranes is on the order of 0.001 µm, which is much smaller than the size of HAB cells, especially when HAB cells are present as colonies consisting of dozens or more individual cells. As a result, the size of HAB cells will be on the order of micrometers, or greater, depending on the colony size. Therefore, HAB cells are effectively removed by both NF and RO processes by a size exclusion mechanism owing to their greater size in comparison to the typical pore sizes of NF and RO systems. A general framework for assessment of the processes controlling the removal of organic contaminants using NF is shown in Figure 7.1.

The molecular weight and typical size dimensions of HAB toxins are shown in Table 7.2. HAB toxins are considerably smaller than HAB cells but still significantly larger than the pores of an RO membrane. The size of HAB toxins, including microcystins, cylindrospermopsin, anatoxin-a, for example, are much larger than the typical pore size for RO processes. The molecular weight of HAB toxins, however, falls within the range of pore sizes for NF membranes. Therefore, it is expected that "tight" NF membranes with MWCO near the smaller pore size of the spectrum will remove all HAB toxins; "loose" NF membranes may not remove some HAB toxins via size exclusion, especially the smaller toxins, such as anatoxin-a or saxitoxin.

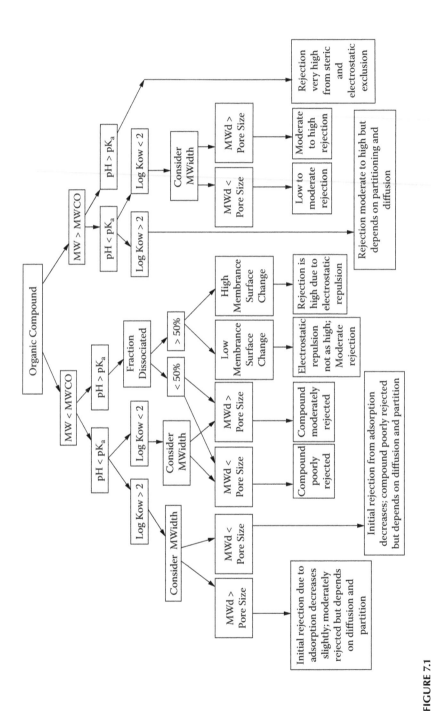

FIGURE 7.1
A flow diagram illustrating the various processes important in the rejection of organic pollutants during membrane treatment. (From C. Bellona, J.E. Drewes, P. Xu, and G. Amy, *Water Research*, 38 (2004) 2795–2809.)

TABLE 7.2

Physical and Chemical Characteristics
of HAB Toxins

Toxin	Molecular Weight
Microcystin-LA	910.06
Microcystin-LF	986.16
Microcystin-LR	995.17
Microcystin-LY	1002.16
Microcystin-RR	1038.20
Microcystin-YR	1045.19
Cylindrospermopsin	415.43
Anatoxin-a	165.23
Saxitoxin	299.29

Teixeira and Rosa examined the removal of microcystin-LR, -LY, and -LF by NF and found good removal (>97%) of this toxin for a variety of experimental conditions [2]. The NF membrane (NFT50) used in their experiments was a thin-film composite made of polypiperazine amide on a polysulfone/polyester support material. The membrane had a pore size of 0.43 nm and an MWCO of 150 Da. As a result, NFT50 can be considered a "tight" NF membrane with pore size and MWCO on the smaller end of the spectrum for NF membranes. The similarly high rejection percentage of all the toxins is expected based on the size of microcystins and size exclusion by the NFT50 membrane. The lowest molecular weight of all the toxins examined was that of microcystin-LF, with a molecular weight of 986.16 Da. Water quality parameters, including natural organic matter (NOM) composition and salt concentration, had little or no impact on microcystin rejection because of the predominance of size exclusion in controlling removal.

More recently, Dixon et al. [3] examined the removal of microcystins by NF, including an NF membrane with a larger pore size in the loose range. Four different microcystin variants were examined: microcystin-LR, -YR, -RR, and -LA. Four different membranes with varying MWCO and materials of construction were included in the study. The MWCO of the membranes varied from 100 to 800 Da. The membranes were made of a variety of materials, including polyamide, sulfonated polyethersulfone, and a combination of polyamide/polysulfone. For microcystin, high toxin rejection was observed for the three membranes with MWCO values of 300 Da or less, which was expected based on the molecular weight of all the microcystins examined. The molecular weights of microcystin-LR, -YR, -RR, and -LA were 995.17, 1045.19, 1038.20, and 910.06 Da, respectively, all significantly greater than 300 Da.

Interestingly, some breakthrough of microcystins was observed by Dixon et al. for the NF membrane (NTR7450) with larger pore sizes. For the four different microcystin variants examined, rejection from source water from both the Palmer and Myponga water treatment plants varied from close to 40% to 100%. The NTR7450 membrane had a reported MWCO in the range of 600–800, but experiments with polyethylene glycol (PEG) standards showed at least some breakthrough of polymers of larger size [3]. For this loose NF membrane, rejection was not directly related to the size of the microcystin variant. For example, microcystin-LA had the greatest rejection of any of the molecules examined despite having the lowest molecular weight. Dixon et al. speculated that the differences in rejection were a result of the molecular configuration of the molecules (e.g., as measured by the surface diffusion coefficient), charge, or hydrophobicity.

Dixon et al. [3] also evaluated the removal of cylindrospermopsin by both loose and tight NF membranes. As discussed, cylindrospermopsin is smaller in size than microcystins; therefore, size exclusion may be a less-important removal mechanism especially for loose membranes. As described, Dixon et al. examined four different membranes with varying MWCOs and construction materials. The MWCO of the membranes varied from 100 to 800. The membranes were made of a variety of materials, including polyamide, sulfonated polyethersulfone, and a combination of polyamide/polysulfone. For cylindrospermopsin, high toxin rejection was observed for the three membranes with MWCO values of 300 Da or less, which was expected based on size exclusion and the size of cylindrospermopsin (415 Da). For the NF membrane (NTR7450) with larger pore size, only about 50% of the toxin was removed in Palmer source water. The NTR7450 NF membrane had an MWCO of 600–800 Da; therefore, significant breakthrough of the toxin was expected. The removal of cylindrospermopsin by NTR7450 was greater, roughly 60–90% depending on the filtration time, using source water from the Myponga drinking water reservoir. The reason for the different rejection from the two source waters (Palmer and Myponga) was unclear. Similar changes in flux in response to NOM were observed for the two different source waters. Therefore, differences in fouling were not believed to be the source of the different rejections of cylindrospermopsin. Other water quality characteristics were also similar between the two reservoirs.

The rejection of anatoxin-a has also been demonstrated for tight NF membranes [4]. Anatoxin-a is one of the smallest HAB toxins, with a molecular weight of 165 Da. Therefore, size exclusion is only expected to be an important mechanism for tight NF membranes. Ribau et al. [4] found near-complete removal of anatoxin-a by the 150-Da MWCO NFT50 NF membrane, regardless of source water characteristics. Unlike other HAB toxins (e.g., microcystin-LR) that are expected to be negatively charged under typical water quality conditions, anatoxin-a has a net positive charge. Therefore, significant electrostatic interaction is expected between the positively charged toxin and the negative charge of most commercial membranes.

Coral et al. [5] examined the rejection of saxitoxin by NF membranes. They examined the effectiveness of two different thin-film, polyamide NF membranes: NF-270 and NF-90. NF-90 had a pore size of 0.68 nm and MWCO of

200 Da and was relatively hydrophobic, with a contact angle of 64%. NF-270 had a slightly larger pore size of 0.83 nm and an MWCO of approximately 400 Da and was more hydrophilic (contact angle 38%). Bench-scale experiments with flat sheet membranes showed 100% rejection of saxitoxin by the NF-90 membrane. For the NF-270 membrane, however, rejection varied as a function of operating time, ranging from 11% to 41%. This last result was consistent with the MWCO of the NF-270 membrane, which is equivalent or slightly larger than the molecular weight of saxitoxin. Also, the more hydrophilic nature of NF-270 would further prevent adsorption and removal of saxitoxin, at least during early stages of membrane operations. In fact, the rejection of saxitoxin by NF-270 decreased with filtration time, consistent with the saturation of the membrane surface with the toxin as the operation progressed.

Although size exclusion is typically the dominant mechanism controlling the rejection of HAB toxins from the permeate of NF and RO systems, significant adsorption of HAB toxins to membrane materials has been observed. As mentioned, anatoxin-a has a net positive charge; therefore, significant electrostatic attraction is expected between the toxin and negatively charged membrane surfaces. Hydrophobic interactions are also expected to play an important role in controlling the adsorption of HAB toxins to membrane materials. Lee and Walker [6] examined the adsorption of microcystin-LR on a variety of membrane materials with varying hydrophobicity and found that increased adsorption roughly correlated with increased membrane hydrophobicity (see Figure 7.2). The greatest adsorption of microcystin-LR, for example, was observed on a hydrophobic polysulfone membrane; the lowest adsorbed amount was seen for the much more hydrophilic cellulose acetate membrane material. In fact, for the hydrophilic membrane material, virtually no adsorption occurred (<1%); for the hydrophobic membrane, almost complete adsorption was noted (>90%), at least at the early stages of treatment until the membrane became saturated with the toxin.

The implications of HAB toxin adsorption on membrane surfaces are significant, even if rejection is controlled by size exclusion. The accumulation of toxins on the surface provides a reservoir of hazardous compounds that can potentially be released during backwashing or other cleaning operations. In fact, significantly higher concentrations of toxins could potentially be released on backwashing as compared to source water concentrations. For example, based on a simple mass balance, the concentration of released toxin could be tens to hundreds of times greater in the backwash water depending on the amount of water processed between backwashes and the extent of toxin release. Also, for loose NF membranes, changes in water chemistry could facilitate the release of compounds and greater toxin breakthrough as a result.

The presence of NOM will also influence the adsorption of HAB toxins, depending on the extent of fouling, the particular toxin, and membrane material of construction. Fouling can have an impact on adsorption and rejection

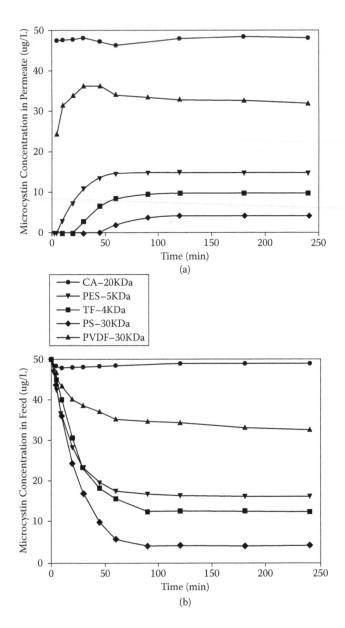

FIGURE 7.2

Concentration of microcystin-LR in permeate flow (a) and feed tank (b) for various membranes. Experimental conditions: microcystin-LR = 50 μg/L, initial permeate flux = 3.87×10^{-5} m^3/(m^2 s), pH 7.0 ± 0.2, ionic strength = 5 mM, and temperature = 23 ± 1°C. (From J. Lee and H.W. Walker, *Journal of Membrane Science*, 320 (2008) 240–247.)

through both physical and chemical mechanisms. When fouling by NOM is significant, for example, the effective pore size of the membranes is reduced, thereby increasing the importance of size exclusion as a rejection mechanism. In such cases, greater rejection of HAB toxins can be expected, even for loose NF membranes. Fouling by NOM will also influence the hydrophobic, electrostatic, and specific chemical interactions between HAB toxins and the membrane surface. Consideration of how NOM competes with HAB toxins for surface sites on the virgin membrane, as well as how toxins may interact directly with NOM deposited on the membrane surface are potentially important to consider. The large number of factors, including NOM heterogeneity, the variety of membrane materials, and varying water chemistry, make it challenging to elucidate clearly the role of NOM in specific cases.

Reverse osmosis is increasingly being used for the desalination of seawater. As a result, there is interest in understanding the ability of RO systems to remove HAB toxins originating not only from freshwater HABs but also from HABs in marine systems. Also, the operational problems introduced by marine HABs, such as membrane fouling, increased chemical consumption, and system disruptions, are of interest [7]. Like freshwater, HABs are an increasing threat to marine and coastal ecosystems [7]. In marine environments, HABs have an impact on recreation and wildlife and threaten fisheries. The harmful algae species *Alexandrium* produces saxitoxin that results in paralytic shellfish poisoning (PSP). The diatom *Pseudo-nitzschia* produces the neurotoxin domoic acid, which causes amnesic shellfish poisoning (ASP). The dinoflagellate *Karenia brevis* produces the so-called brevetoxins that result in neurotoxic shellfish poisoning (NSP) in humans. Other HABs of concern in marine systems include *A. anophagefferens*, *Lingulodinium polyedrum*, and others.

Domoic acid has a molecular weight of 311 g/mol and therefore is in the size range of toxins familiar in freshwater systems. Brevetoxins are a class of HAB toxins with a variety of congeners of different molecular weight. However, these molecules are relatively large, comparable in size to some of the larger freshwater HAB toxins. Brevetoxins have a molecular weight in the range of 900 g/mol. The molecular weight of brevetoxin 2, for example, is 895 g/mol. Laycock et al. [8] examined the removal of marine HAB toxins using RO and found greater than 99% removal for domoic acid and saxitoxins. Seubert et al. [9] examined the removal of marine HAB toxins using RO in both bench-scale and pilot plant systems. In the bench-scale studies, they found that saxitoxins, brevetoxins, and domoic acid were all effectively removed by RO to levels below the limit of detection. In studies carried out at a pilot-scale RO plant in El Segundo, California, saxitoxin and domoic acid were periodically detected in the source water to the facility. However, these toxins were never detected in the permeate stream, indicating complete removal in the RO process.

7.2.2 Ultrafiltration and Microfiltration

Fouling by NOM can be especially severe for the treatment of surface water by RO and NF membranes and usually requires some form of pretreatment, such as MF or UF. As a result, a significant amount of research has been carried out examining the performance of UF and MF for the removal of HAB cells and toxins. Unlike RO and NF, the typical pore sizes of MF and UF are orders of magnitude larger than the size of most HAB toxins. Therefore, size exclusion is not expected to be an important toxin rejection mechanism for these types of membrane systems. The typical pore sizes of UF and MF membranes, however, are small enough to effectively reject most HAB cells, especially when present in colonies.

The rejection of HAB cells by UF and MF is well documented in the research literature. Gijsbertsen-Abrahamse et al. [10] examined the removal of *Planktothrix agardhii* and *Planktothrix rubescens* using a lab-scale UF system. The membrane used in these experiments was made of polyethersulfone and polyvinylpyrrolidone in a hollow-fiber configuration with an MWCO of 100 kDa and an average pore diameter of 30 μm. Performance tests demonstrated greater than 99.99% removal of microorganisms. Microscopic analysis showed that the average length of *P. agardhii* filaments was approximately 200 μm, therefore significantly larger than the average pore size of the hollow-fiber membranes.

Campinas and Rosa [11] examined the removal of *Microcystis* and the potential for toxin release in a lab-scale, hollow-fiber membrane system. The membrane was made of cellulose acetate and had an MWCO of 100 kDa. Similar to Gijsbertsen-Abrahamse, they found essentially complete removal of HAB cells for the duration of their experiments and under all experimental conditions. Although they found some cell lysis occurred, the permeate quality never deteriorated in terms of the concentration of toxin compared to the influent feed concentration. Interestingly, Campinas and Rosa also explored the potential release of microcystins at different growth phases. They found that cell lysis occurred at all growth phases but was slightly more pronounced for older cells. Little or no adsorption of microcystins was observed on the membrane, consistent with other studies [6].

Lee and Walker examined the rejection of microcystin-LR using a number of different UF membranes, including polyamide, cellulose acetate, polyvinylidene fluoride, polyethersulfone, and polysulfone [6]. As described previously, they found that the membrane material played an important role in controlling the extent of microcystin adsorption, with adsorption on the membrane increasing with increasing membrane hydrophobicity. They also found some removal of microcystin-LR via a size exclusion mechanism for a tight thin-film composite membrane with an MWCO of 1000 Da. For membranes with an MWCO higher than 2000 Da, little or no removal by size exclusion was observed, consistent with the known molecular weight of microcystin-LR.

Although a number of studies have been carried out to examine the removal of HAB cells and toxins by UF and MF at the lab scale, comparatively fewer have documented removal at the pilot or full scale. Sorlini et al. [12] showed effective removal of HAB cells, including *Anabaena* sp. and *Microcystis* sp., with removal percentages of 98% or higher in a hollow-fiber MF pilot plant study. The MF membrane was made of polyvinylidene and had a pore size of 0.1 μm and an MWCO of 200 kDa. Therefore, these data are consistent with previous lab-scale studies and further support the use of MF and UF for effective removal of HAB cells for drinking water treatment.

Although UF and MF have proven effective at removing HAB cells, concern has been raised regarding the potential release of intercellular toxins during membrane operations. Most full-scale membrane plants operate in cross-flow mode, which has the potential of creating shear forces that may lead to cell lysis and toxin release. Also, concern exists about the potential for cell disruption and toxin release during pumping. A few studies have examined the potential for toxin release during MF and UF operations. Gijsbertsen-Abrahamse et al. [10] examined the release of HAB toxin from *P. agardhii* and *P. rubescens* cells during hollow-fiber membrane filtration. They found low toxin release. In fact, less than 2% of cell-bound microcystin was released over the course of their experiments. In a pilot plant study, Sorlini et al. [12] showed little or no release of intercellular toxin in a hollow-fiber MF pilot plant. Additional study, however, is needed to more fully understand the conditions that may lead to toxin release in UF and MF systems.

7.2.3 Ultrafiltration–Powdered Activated Carbon

Ultrafiltration and MF are generally not effective processes for the removal of HAB toxins. However, UF and MF can be coupled to an adsorption process to achieve simultaneous solid and toxin removal. For example, UF has been coupled with PAC for the removal of taste and odor compounds, herbicides, pesticides, and other hazardous trace organic compounds [13]. The use of PAC-UF has also been explored for the simultaneous removal of HAB cells and toxins. In one of the first studies, Lee and Walker [14] showed effective removal of microcystin-LR using a hybrid PAC-UF system at the bench scale. They demonstrated that effective removal of microcystin-LR depended on both the type of carbon used and the membrane material and pore size.

As discussed previously, the effectiveness of PAC for HAB toxin removal depends critically on the amount of mesopore volume. Donati et al. [15], for example, showed that the adsorption density of microcystin-LR on eight different activated carbons correlated with the amount of mesopore volume and was unrelated to the micropore volume. The significance of mesopore volume can be understood by comparing the size of HAB toxins relative to the diameter of mesopores. Mesopores are characterized as pores with diameters in the range of 2 to 50 nm; micropores are pores less than 2 nm. The size of microcystin-LR, for example, has been reported to be from 1.2 to

2.6 nm [15]. Therefore, microcystin-LR molecules may access adsorption sites within mesopores in the activated carbon, but not micropores.

Activated carbon produced from coconut shells generally has the greatest amount of micropore volume; wood- and coal-based carbons have greater mesopore and macropore volume [13]. The distribution of pore size in active carbon is typically attributed to the inherent physical structure of the raw material [13], with coconut shell processing properties that result in mostly micropore volume during production. Although micropore volume is desirable for removal of most micropollutants, the relatively large size of HAB toxins like microcystins makes microporous carbon less effective. As a result, greater adsorption of microcystin-LR is observed with wood-based carbon compared to coconut-based materials [14, 15]. Saxitoxins, on the other hand, are more effectively removed by activated carbon with greater micropore volume [16] owing to the smaller size of saxitoxins.

In the study by Lee and Walker [14], greater removal of microcystin-LR was achieved using a wood-based carbon, consistent with the studies of PAC treatment of HAB toxins described previously in this chapter. In fact, wood-based carbon was four times more effective at removing microcystin-LR compared to a coconut-based carbon. Lee and Walker also showed that effective removal of microcystin-LR was achieved without significant adsorption to the membrane material, at least in the case of cellulose acetate membranes. For more hydrophobic membranes, the amount of microcystin-LR removal was greater owing to the additional contribution because of adsorption of toxin on the membrane surface. However, the additional removal of toxin was only temporary until the adsorption capacity of the membrane material was reached.

Like other PAC processes, the presence of NOM has the potential to reduce the capacity of carbon for toxin removal. Lee and Walker [14], for example, showed that the presence of Suwanee River fulvic acid reduced the removal of microcystin-LR by PAC-UF because of competition between the toxin and fulvic acid for sites on the adsorbent surface. Campinas and Rosa [17] also examined the removal of microcystin-LR by PAC-UF and confirmed the effectiveness of this approach. They found that a PAC dosage of 15 mg/L could effectively treat source water containing 20 µg/L of microcystin-LR to below the World Health Organization (WHO) drinking water guideline. In the presence of 5 mg/L NOM, however, PAC-UF was less effective at removing microcystin-LR, and it was not possible to consistently meet the WHO guideline. Zhang et al. [18] showed near-complete removal of *Microcystis* cells using PAC-UF, but only partial removal of microcystin-LR with permeate concentrations between 3 and 5 µg/L.

PAC-UF has also been coupled with coagulation for improved HAB cell and toxin removal. Coupling coagulation with PAC-UF potentially improves turbidity removal and reduces membrane fouling. As discussed in the previous chapter, metal salt coagulation has little or no effect on dissolved toxins. Dixon et al. [19], for example, examined a coupled coagulation-PAC-UF system for the removal of *Microcystis* cells and microcystin-LR. They found that UF, both with and without coagulation, effectively removed *Microcystis* cells, and PAC reduced microcystin concentrations. The addition of coagulation reduced the potential for membrane fouling and improved flux over the course of the experimental runs. Dixon et al. [20] also explored a similar system for *Anabaena* sp. and saxitoxin removal and found similar qualitative results.

7.3 Advanced Oxidation Processes

Advanced oxidation processes represent another class of emerging technologies increasingly investigated to deal with HAB cells and toxins. In addition to more conventional oxidation technologies, AOPs show promise for destroying HAB toxins in drinking water. AOPs currently being developed and examined for destroying HAB toxins include photolysis, UV/H_2O_2, UV/TiO_2, technologies based on generation of sulfate radical, ultrasound, Fenton reaction, and ferrate (for a review of AOPs for HABs and HAB toxins, see [21]). This section provides an introduction to oxidation technologies and their application to removing or destroying HAB toxins in drinking water.

7.3.1 Ozone

Ozone is an oxidant commonly used in water treatment for the disinfection of disease-causing organisms. Ozone is increasingly being used as an oxidation process to destroy organic contaminants. Ozone destroys contaminants during water treatment applications by two distinct mechanisms: direct attack by molecular ozone or indirect reactions as a result of the decomposition of ozone and subsequent production of hydroxyl radical [22]. Research has shown that ozone is effective for destroying a number of HAB toxins, except saxitoxins. Like the application of chlorine discussed in the previous chapter, there are two important questions to consider when using ozone for the control of HAB cells and HAB toxins in drinking water: (1) What is the required dose and contact time needed to sufficiently destroy specific HAB toxins? and (2) To what extent will disinfection disrupt HAB cells and lead to the release of intercellular toxin into the treated water?

Numerous studies have examined the effect of ozone on the destruction of microcystins in drinking water (for reviews, see [23–25]). Effective destruction of microcystin by ozone has been shown to occur on the timescale of seconds or minutes. Shawwa et al. [26], for example, showed that the destruction of microcystin-LR by ozone was rapid and could be represented as a second-order reaction (first order with respect to both ozone concentration and microcystin-LR concentration). They also showed that the rate constant increased with temperature, from 10°C to 30°C, and the presence of NOM resulted in an initial lag in degradation. An early study by Rositano et al.

[27] demonstrated that the destruction of microcystin-LR was pH dependent, with less-effective destruction under alkaline pH conditions because of the lower oxidation potential of ozone under these conditions. Rositano et al. [27] also showed that ozone is the most effective oxidant for the destruction of microcystin, with a reaction rate faster than hydrogen peroxide, chlorine, and permanganate.

The mechanism of microcystin breakdown by ozone has been an area of research in recent years, not only to understand the degradation process but also to identify possible toxic degradation by-products. In one study by Miao et al. [28], the intermediate by-products in the destruction of microcystin-LR and microcystin-RR by ozone were identified using liquid chromatography/ mass spectrometry (LC-MS). This study found that attack on the Adda chain by hydroxyl radical and molecular ozone was the most important degradation pathway for both microcystin-LR and microcystin-RR. They also observed cleavage of the peptide ring by ozone but suggested this was not as important a degradation pathway as attack on the Adda side chain. Miao et al. also demonstrated that the toxicity of the resulting by-products was significantly reduced compared to the parent compounds. Brooke et al. [29] found the use of ozone destroyed microcystins and decreased toxicity.

The use of ozone to destroy cylindrospermopsin is also well documented in the research literature. For example, 95% destruction of cylindrospermopsin has been documented at an ozone dosage of 0.38 mg/L in static-dose testing [24]. Similar to microcystins, the effective destruction of cylindrospermopsin by ozone occurs on the timescale of seconds or minutes. In fact, Rodriguez et al. estimated the half-life of cylindrospermopsin as 0.1 second at a constant ozone residual concentration of 1.0 mg/L. Also, similar to microcystin, the destruction of cylindrospermopsin is typically represented as a second-order reaction and is dependent on pH [25]. For cylindrospermopsin, ozone demonstrates the fastest kinetic breakdown compared to chlorine and permanganate. The degradation by-products from the reaction of ozone with cylindrospermopsin are less well understood compared to microcystins.

Ozone also effectively destroys anatoxin-a. An ozone dosage of 0.75 mg/L resulted in 95% destruction of anatoxin-a [24]. The effective destruction of anatoxin-a by ozone occurs on the timescale of seconds or minutes. Rodriguez et al. estimated the half-life of anatoxin-a as 0.52 second at a constant ozone residual concentration of 1.0 mg/L. For anatoxin-a, ozone demonstrated faster kinetics compared to chlorine but was not as fast as permanganate [24]. Although the destruction of anatoxin-a by ozone has been described, the degradation by-products arising from the reaction of ozone with anatoxin-a are not well understood. Rositano et al. [30] showed near-complete destruction of anatoxin-a in four Australian waters at an ozone concentration of 1.5 mg/L. Below this concentration, the extent of removal depended on the particular water. Anatoxin-a in Hope Valley water, for example, required an ozone dosage of just less than 1 mg/L for 90% removal; an equivalent amount of toxin removal in Edenhope water required an ozone dosage closer to 1.5 mg/L.

Studies with ozone showed that saxitoxins are largely resistant to degradation. Orr et al. examined the destruction of a number of saxitoxins, including saxitoxin (STX), gonyautoxin (GTX-5), and decarbamoylsaxitoxin (dc-STX), in a lab-scale experimental system using ozone concentrations as high as 2 mg/L. They found little or no destruction of any of the saxitoxins [31]. As expected, little reduction in solution toxicity was also observed. Orr et al. also showed ozone combined with hydrogen peroxide was ineffective in destroying saxitoxin. Rositano et al. [30] examined the destruction of different saxitoxins (i.e., GTX2, GTX3, C1, and C2) in four Australian waters. Like Orr et al., they also showed little or no destruction of saxitoxin at ozone concentrations of 2 mg/L or less. Rositano et al. did observe some degradation of saxitoxin for ozone concentrations greater than 2 mg/L. At ozone concentrations greater than 2 mg/L, 25–50% saxitoxin removal was observed.

It should be noted that the application of ozone can lead to high levels of bromate in the treated water. Bromate is formed in drinking water when ozone reacts with the bromide ion. Bromate is a suspected human carcinogen, and the United States Environmental Protection Agency has established a maximum contaminant level (MCL) in drinking water of 10 mg/L. Rodriguez et al. [24] found that ozone dosages of 0.25 mg/L resulted in high microcystin destruction (>95%) and low bromate formation. At higher ozone concentrations (>1 mg/L), however, bromate formation became significant (>1 mg/L) but did not exceed the MCL value.

The rate of HAB toxin destruction by ozone is often represented in terms of an apparent rate constant, in which

$$C = C_0 e^{-k_{app} \times dose \times t}$$

where C is the final HAB toxin concentration, C_0 is the initial concentration, *dose* is the dosage of ozone, and k_{app} is the apparent rate constant. This equation can be used to determine the required contact time (or dosage) needed to achieve a certain level of HAB toxin destruction. For example, consider the contact time needed to reduce microcystin-LR from 50 to 1 µg/L at a pH of 8 and temperature of 20°C using ozone. The apparent rate constant k_{app} for the reaction of ozone with microcystin-LR at pH 8 and a temperature of 20°C is 4.1×10^5 M^{-1} s^{-1} [24]. Assuming an ozone dosage of 0.50 mg/L, the time required for the reaction is

$$t = -\frac{ln\left(C/C_0\right)}{k_{app} \times dose}$$

$$t = -\frac{ln\left(1/50\right)}{4.1 \times 10^5 \times 0.50} \times 800 = 1.5 \times 10^{-2} \text{ minutes}$$

The calculated value for t in these equations typically represents t_{10} to provide a conservative estimate of the flow in the reaction. Therefore, the average hydraulic detention time t_d, considering a baffling factor of 0.7, would be

$$t_d = t/0.7 = \frac{1.5 \times 10^{-2}}{0.7} \times \frac{60 \text{ sec}}{\text{min}} = 1.3 \text{ sec}.$$

This calculation indicates approximately 1.3 seconds of reaction time is needed to reduce the microcystin-LR concentration from 50 to 1 µg/L, assuming an ozone dosage of 0.50 mg/L.

7.3.2 Potassium Permanganate

Permanganate has seen widespread application as a chemical for water treatment. It is used for a variety of purposes, including the oxidation of taste and odor compounds, oxidation of iron and manganese, and control of nuisance organisms (e.g., zebra mussels) [13]. Permanganate has also seen considerable use in water treatment as a "preoxidant" to oxidize precursors to the formation of disinfection by-products (DBPs) during chlorination [13]. Permanganate is typically added to water as the salt, potassium permanganate. Under acidic conditions,

$$MnO_4^- + 4H^+ + 3e^- \rightarrow MnO_{2(s)} = 2H_2O$$

and under basic conditions,

$$MnO_4^- + 2H_2O + 3e^- \rightarrow MnO_{2(s)} = 4OH^-$$

One of the first studies using potassium permanganate to destroy microcystin was carried out by Chen et al. [32]. They found permanganate was effective in destroying microcystin-RR with an estimated half-life, based on a second-order reaction mechanism, of less than 1 min for $KMnO_4$ concentrations between 5 and 10 mg/L. Near-complete removal (99.5%) was observed after 10 minutes. Chen et al. showed that the rate constant for the reaction depended only slightly on solution pH. For a $KMnO_4$ concentration of 5 mg/L and a pH of 5, for example, the half-life for the reaction was 0.59 minutes. When the pH increased to 9.0, the reaction was just slightly slower, with a calculated half-life of 75 minutes.

Rodriguez et al. [24, 33] also examined the use of permanganate for the destruction of microcystins. They found similar rate constants to Chen et al. [32] for microcystin-LR, -YR, and -RR and little influence of solution pH. Rodriguez et al. estimated a $KMnO_4$ dosage of 1–1.25 mg/L was generally sufficient to meet the WHO limit for microcystins of 1 µg/L. Based on the similarity in rate constants between the different microcystin variants, Rodriguez et al. suggested that permanganate reactions occurred with the Adda moiety on the microcystin molecule. In comparing second-order rate constants, they concluded permanganate degraded microcystins faster than chlorine and chlorine dioxide, but slower than ozone.

Similar studies have been carried out to examine the degradation of cylindrospermopsin and anatoxin-a by potassium permanganate. Rodriguez et al. [34] showed that $KMnO_4$ is largely ineffective at degrading cylindrospermopsin and not a viable oxidant for water treatment applications. Anatoxin-a, on the other hand, is quickly destroyed by potassium permanganate, faster in fact than microcystin-LR [24, 34]. Permanganate is a more effective oxidant than ozone or chlorine for destroying anatoxin-a. A comparison of various oxidants for destroying HAB toxins is shown in Figure 7.3.

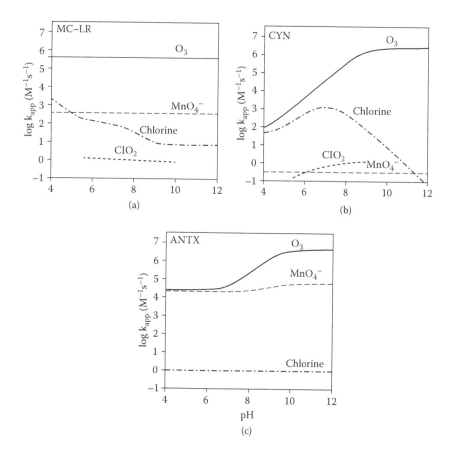

FIGURE 7.3
Effect of pH on reaction of HAB toxins with various chemical oxidants. (From E. Rodríguez, G.D. Onstad, T.P.J. Kull, J.S. Metcalf, J.L. Acero, and U. von Gunten, *Water Research*, 41 (2007) 3381–3393.)

7.3.3 Photolysis and UV/Hydrogen Peroxide

Photolysis refers to the breakdown of a chemical compound in the presence of photons. Photon energy in the visible, UV, x-ray, and gamma-ray spectrum have all been used in photolysis reactions. Chemical compounds may directly absorb photons, or the photon energy may be transferred to the compound via a photosensitizer. Photolysis occurs in both natural and engineered systems. In pure water, microcystins [35], cylindrospermopsin [36], anatoxin-a, and saxitoxin break down slowly in the presence of sunlight, but photosensitizers (e.g., algal pigments, humic or fulvic acid) may accelerate the breakdown in natural systems. Given the potential for the breakdown of HAB toxins by photolysis and related processes, many studies have been carried out to develop this technology for drinking water applications.

The photolytic breakdown of microcystin by UV light has been documented [37]. At an intensity of 147 $\mu W/cm^2$, the half-life of microcystin was approximately 10 minutes. These results are similar to a study by Qiao et al. [38] that showed a half-life for microcystin-RR, at a comparable light intensity of 3.66 mW/cm^2, of about 29 minutes. The difference in half-lives probably reflects the different photochemical reactor used as well as water quality conditions. At a higher UV intensity of 2550 mW/cm^2, Tsuji et al. [37] showed that microcystin was completely decomposed after 10 minutes.

The combination of UV with hydrogen peroxide has been shown to increase the rate and extent of microcystin degradation, compared to the use of UV or peroxide individually [38]. In the presence of hydrogen peroxide alone, little or no breakdown of microcystin was observed. In the presence of UV light, hydrogen peroxide decomposes to form hydroxyl radical, according to

$$H_2O_2 \xrightarrow{h\upsilon} 2 \ ^{\bullet}OH$$

The combined effect of direct photolysis and decomposition by the hydroxyl radical increases the extent of microcystin breakdown. For a UV light intensity of 3.66 mW/cm^2, the half-life of microcystin-RR in the presence of UV/H_2O_2 was approximately 10 minutes, and microcystin-RR was almost completely destroyed after 60 minutes. By comparison, the half-life of microcystin-RR on exposure to UV light alone was closer to 20 minutes. Less than 10% of the microcystin-RR was degraded by H_2O_2 alone, even after 60 minutes. As mentioned, the greater degradation of microcystin by UV/H_2O_2 is proposed to be because of the combined action of direct photolysis and decomposition by hydroxyl radical. The reaction pathways for the breakdown of microcystin-LR by hydroxyl radical are shown in Figures 7.4 and 7.5. Song et al. suggested that the hydroxyl radical reacts primarily with benzene and diene functional groups on the Adda chain [39].

Chiswell et al. examined the photolysis of cylindrospermopsin in the presence of UV light [36]. At 400 mW/m^2, they found the half-life of cylindrospermopsin was 18 hours. Cheng et al. [40] examined the UV photolysis of

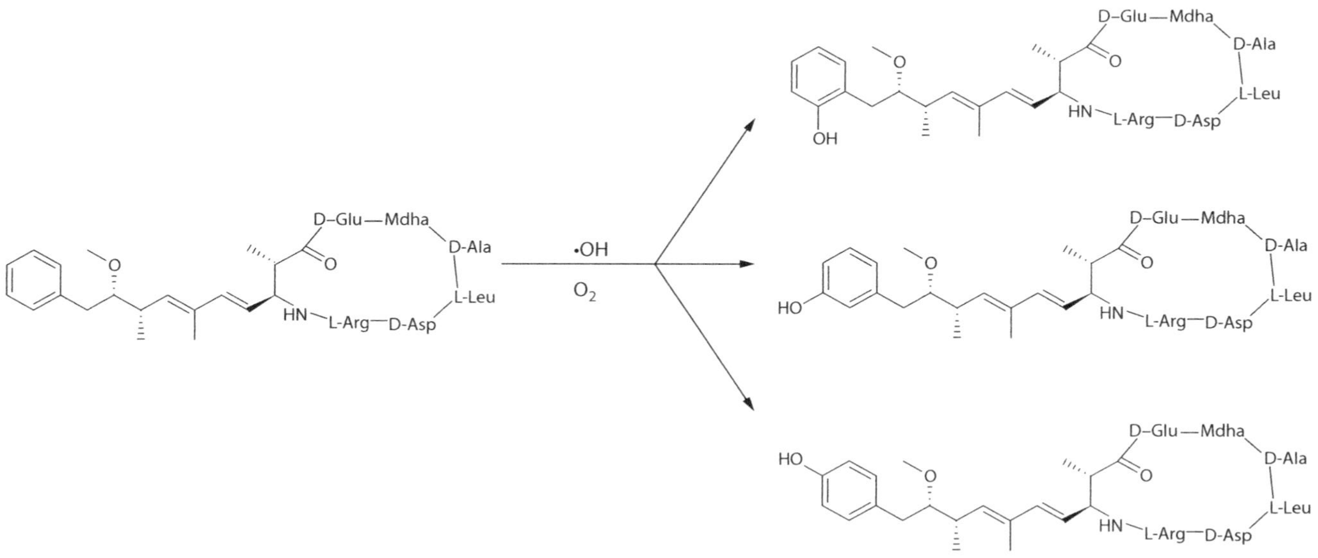

FIGURE 7.4

Reaction of hydroxyl radical with the benzene group in microcystin-LR. (From W. Song, T. Xu, W.J. Cooper, D.D. Dionysiou, A.A.d.l. Cruz, and K.E. O'Shea, *Environmental Science and Technology*, 43 (2009) 1487–1492.)

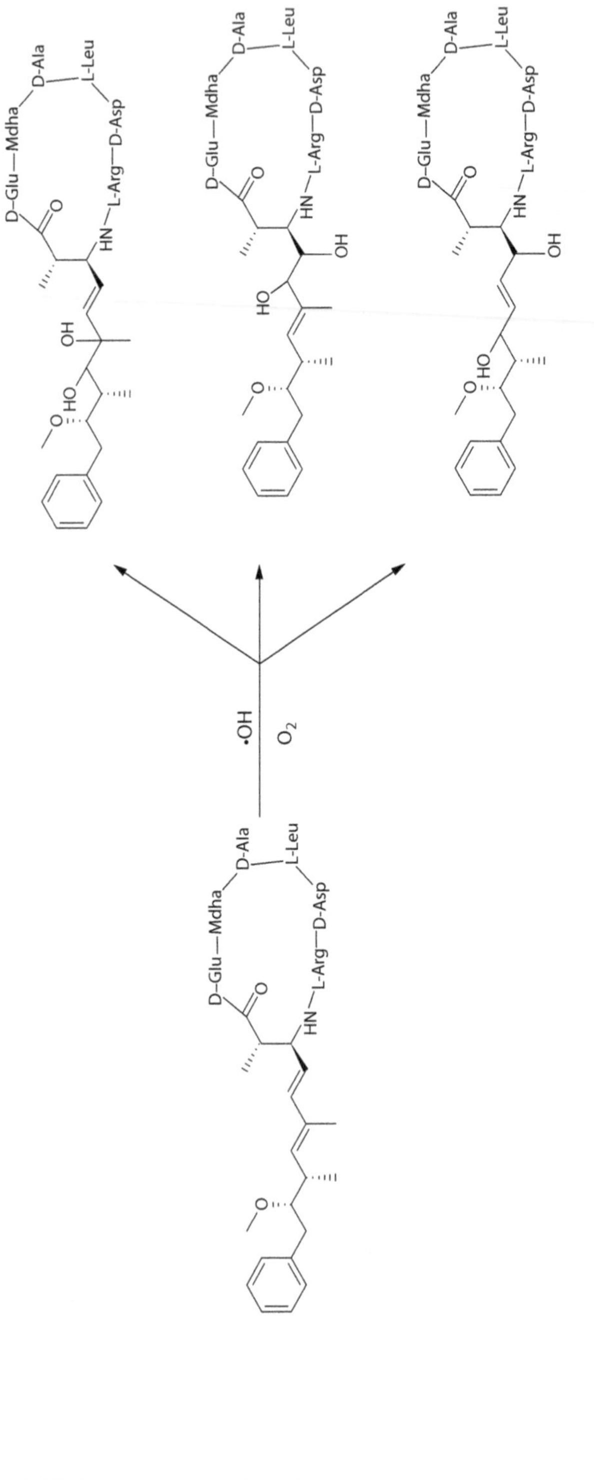

FIGURE 7.5
Reaction of hydroxyl radical with the diene group in microcystin-LR. (From W. Song, T. Xu, W.J. Cooper, D.D. Dionysiou, A.A.d.l. Cruz, and K.E. O'Shea, *Environmental Science and Technology*, 43 (2009) 1487–1492.)

cylindrospermopsin and concluded that very high UV doses are needed for effective decomposition, doses that are much higher than would be typically used in a drinking water treatment plant. They also showed that UV light alone was not effective at inactivating *Cylindrospermopsis raciborskii* at typical UV dosages used in practice.

The UV/H_2O_2 AOP has been shown to destroy cylindrospermopsin more effectively than in the presence of UV light alone. He et al. [41] demonstrated the rapid decomposition of cylindrospermopsin using H_2O_2 in the presence of UV light at 254 nm. Over 100 potential reaction by-products were identified by MS and represented three reaction pathways: hydroxyl addition, alcoholic oxidation or dehydrogenation, and elimination of sulfate. In a separate study, it was shown that hydroxyl radicals react primarily with the uracil ring of cylindrospermopsin [42]. The reaction pathway for the breakdown of cylindrospermopsin by hydroxyl radical is shown in Figure 7.6.

Effective degradation of anatoxin-a has been demonstrated in the presence of UV light and H_2O_2 [43]. Afzal et al. [43] looked at the destruction of anatoxin-a using a low-pressure UV lamp (200 mJ/cm²), and an H_2O_2 concentration of 30 mg/L resulted in 70% degradation. In the absence of hydrogen peroxide, much higher UV dosages were needed to see any significant removal. A six-times-higher dosage of 1285 mJ/cm² was required, for example, to degrade a 1.8-mg/L anatoxin-a solution by 50% in the absence of hydrogen peroxide. Little is known about the destruction of saxitoxin in the presence of UV light with or without hydrogen peroxide.

7.3.4 UV/TiO₂

The photoexcitation of a TiO_2 catalyst generates hydroxyl radical (OH·) and superoxide radical ($O_2^{\cdot-}$) capable of destroying broad classes of contaminants in water [44, 45]. A schematic of the process is shown in Figure 7.7. A number of studies have examined the application of the UV/TiO_2 process for the destruction of HAB toxins in drinking water (for a review, see Sharma et al. [23]). Shephard et al. [46], for example, examined the destruction of microcystin-LR, -YR, and -YA using a "falling film" photochemical reactor and found effective destruction of all the toxins. The half-life for the degradation was generally less than 5 minutes, although the time required increased in natural water (compared to distilled) and at lower catalyst loading rates. Lawton et al. [47] also demonstrated the effectiveness of UV/TiO_2 for the destruction of microcystins and found that the process depended on pH and the amino acid composition of the toxins. Maximum degradation rates for microcystin-LR were observed at pH values between 4 and 7; microcystin-LF had high rates of destruction at pH values from 5 to 12. The influence of pH on photocatalytic destruction was suggested to be caused by changes in the charge and hydrophobicity of the toxins and catalyst at different pH values. The UV/TiO_2 process has also been shown to effectively degrade cylindrospermopsin [48]. They found that the degradation rate of cylindrospermopsin

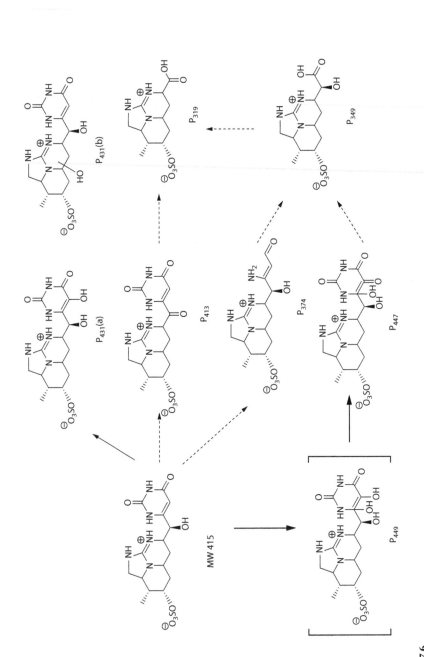

FIGURE 7.6
Reaction of hydroxyl radical with cylindrospermopsin. (From W. Song, S. Yan, W.J. Cooper, D.D. Dionysiou, and K.E. O'Shea, *Environmental Science and Technology*, 46 (2012) 12608–12615.)

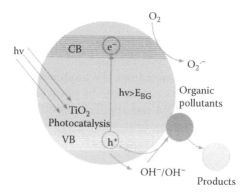

FIGURE 7.7

The mechanism by which organic pollutants are destroyed during TiO_2 photoexcitation in the presence of oxygen. (From V.K. Sharma, T.M. Triantis, M.G. Antoniou, X. He, M. Pelaez, C. Han, W. Song, K.E. O'Shea, A.A. de la Cruz, T. Kaloudis, A. Hiskia, and D.D. Dionysiou, *Separation and Purification Technology*, 91 (2012) 3–17.)

depended on both pH and the source of the TiO_2, with higher degradation rates at higher pH values. Tominaga et al. [49] showed effective degradation of saxitoxin using UV light and a surface-modified TiO_2.

7.3.5 Sulfate Radical

The development and application of sulfate radical as an AOP is receiving increasing attention. Sulfate radical AOPs may utilize UV light, thermolysis, or electron transfer processes to "activate" persulfate ($S_2O_8^{2-}$) or peroxymono-sulfate (HSO_5^-) and create sulfate radicals. In the presence of UV light, for example, sulfate radical is formed via the following reactions:

$$S_2O_8^{2-} + h\upsilon \rightarrow SO_4^{\bullet-} + SO_4^{\bullet-}$$

$$HSO_5^- + h\upsilon \rightarrow SO_4^{\bullet-} + {}^{\bullet}OH$$

Sulfate radical is a strong oxidant, stronger than typical water treatment chemicals such as permanganate and hypochlorous acid, with comparable redox potentials to hydroxyl radical. One advantage of sulfate radical, compared to hydroxyl radical, is that sulfate radical is more selective in reacting with unsaturated bonds or aromatic groups, thereby making it a more efficient process for some organic contaminants.

Antoniou et al. [50] examined the destruction of microcystin-LR using sulfate radical and demonstrated this technique was capable of degrading microcystin-LR at rates similar to or greater than other AOPs. In the presence of UV light, the destruction of microcystin-LR depended on the oxidant/microcystin ratio and pH. The rate of degradation increased, for example, with decreasing pH and increasing oxidant/microcystin ratio. He et al. [51] examined the destruction of cylindrospermopsin using UV-activated persulfate and peroxymonosulfate and found effective destruction of this cyanotoxin at rates greater than UV alone and UV/H_2O_2. Persulfate resulted in the fastest degradation, followed by peroxymonosulfate. Hydroxyl radical scavengers, such as NOM and alkalinity, hindered the degradation of cylindrospermopsin by sulfate radical, and the presence of transition metals (e.g., Cu^{2+} and Fe^{2+}) enhanced reaction rates.

7.3.6 Ultrasound

Ultrasound is another AOP that has been investigated as a possible technology for advanced water treatment [52]. Some recent studies focused on the destruction of cyanotoxins in drinking water. The details of the ultrasonic process were reviewed extensively elsewhere [53]. The ultrasonic irradiation of water results in acoustic cavitation and the formation and collapse of microbubbles. The collapse of these bubbles is capable of generating extreme temperatures and pressures that generate a variety of radical species, including hydroxyl radical, hydrogen radicals, and others. The formation of these radical species leads to the destruction of a variety of organic contaminants.

Song et al. [54, 55] examined the use of ultrasound for the destruction of microcystins from drinking water. They found that ultrasound could rapidly degrade both microcystin-LR and microcystin-RR at a frequency of 640 kHz. They determined that reaction with hydroxyl radical was the dominant degradation mechanism. Apparently, the hydroxyl radical reacted with the benzene ring and diene of the Adda moiety. Song et al. also speculated that hydroxyl radical cleaved the peptide ring by reacting at the site of the Mdha-Ala peptide bond. The ultrasound process was more effective at lower pH, possibly because of the greater hydrophobicity of microcystin under acidic conditions. Song et al. also demonstrated that the ultrasonic degradation of microcystin decreased the toxicity of the solution, suggesting the resulting by-products of the ultrasound process were nontoxic. Studies have yet to be carried out examining the destruction of other cyanobacterial toxins, such as anatoxin-a, cylindrospermopsin, and saxitoxin, by ultrasonic processes.

7.3.7 Fenton Reaction

The Fenton reaction has also been explored as an AOP for the destruction of HAB toxins. The reaction of ferrous iron (Fe^{2+}) with hydrogen peroxide results in the generation of hydroxyl radical according to

$$Fe^{2+} + H_2O_2 \rightarrow Fe^{3+} + OH^- + {}^{\bullet}OH.$$

A number of studies have examined the use of the Fenton reaction for the destruction of microcystins [56–58]. In one of the first studies, Gajdek et al. [56] showed the complete degradation of microcystin-LR within 30 minutes using the Fenton reaction. Bandala et al. [57] showed that relatively high peroxide concentrations are needed for effective destruction of microcystin-LR based on the Fenton reaction and that the process can be more effective if combined with UV light (so-called photo-Fenton reaction). Zhong et al. [58] demonstrated that the Fenton reaction could be used to destroy microcystin-RR in water. Under optimal conditions, they showed greater than 99% destruction of the toxin. Al Momani et al. [59] showed the Fenton reaction was capable of destroying anatoxin-a in 90 seconds.

7.3.8 Ferrate

Recently, a dual oxidant-coagulant was examined for the removal of HAB toxins from drinking water. Yuan et al. [60] examined the removal of microcystin-LR using ferrate and found effective removal of this toxin over the pH range of 6–10. Fe(VI) in ferrate decomposes to Fe(III) in water; therefore, ferrate is a powerful oxidant. In fact, Yuan et al. observed 100% removal of the microcystin-LR at a dosage of 40 mg/L. The removal of microcystin was pH dependent, with greater than 80% removal observed over the pH range of 6–10. Once oxidized to Fe(III), the iron precipitates as a hydroxide, thereby serving as a coagulant chemical. Ferrate has not been demonstrated for the destruction of cylindrospermopsin, anatoxin-a, or saxitoxins.

References

1. C. Bellona, J.E. Drewes, P. Xu, and G. Amy, Factors affecting the rejection of organic solutes during NF/RO treatment—a literature review. *Water Research*, 38 (2004) 2795–2809.
2. M.R. Teixeira and M.J. Rosa, Microcystins removal by nanofiltration membranes. *Separation and Purification Technology*, 46 (2005) 192–201.
3. M.B. Dixon, C. Falconet, L. Ho, C.W.K. Chow, B.K. O'Neill, and G. Newcombe, Removal of cyanobacterial metabolites by nanofiltration from two treated waters. *Journal of Hazardous Materials*, 188 (2011) 288–295.
4. M. Ribau Teixeira and M.J. Rosa, Neurotoxic and hepatotoxic cyanotoxins removal by nanofiltration. *Water Research*, 40 (2006) 2837–2846.
5. L.A. Coral, L.A. de Oliveira Proença, F.d.J. Bassetti, and F.R. Lapolli, Nanofiltration membranes applied to the removal of saxitoxin and congeners. *Desalination and Water Treatment*, 27 (2011) 8–17.
6. J. Lee and H.W. Walker, Mechanisms and factors influencing the removal of microcystin-LR by ultrafiltration membranes. *Journal of Membrane Science*, 320 (2008) 240–247.

7. D.A. Caron, M.-È. Garneau, E. Seubert, M.D.A. Howard, L. Darjany, A. Schnetzer, I. Cetinić, G. Filteau, P. Lauri, B. Jones, and S. Trussell, Harmful algae and their potential impacts on desalination operations off southern California. *Water Research*, 44 (2010) 385–416.

8. M.V. Laycock, D.M. Anderson, J. Naar, A. Goodman, D.J. Easy, M.A. Donovan, A. Li, M.A. Quilliam, E. Al Jamali, and R. Alshihi. Laboratory desalination experiments with some algal toxins. *Desalination*, 293 (2012) 1–6.

9. E.L. Seubert, S. Trussell, J. Eagleton, A. Schnetzer, I. Cetinić, P. Lauri, B.H. Jones, and D.A. Caron, Algal toxins and reverse osmosis desalination operations: Laboratory bench testing and field monitoring of domoic acid, saxitoxin, brevetoxin and okadaic acid. *Water Research*, 46 (2012) 6563–6573.

10. A.J. Gijsbertsen-Abrahamse, W. Schmidt, I. Chorus, and S.G.J. Heijman, Removal of cyanotoxins by ultrafiltration and nanofiltration. *Journal of Membrane Science*, 276 (2006) 252–259.

11. M. Campinas and M.J. Rosa, Evaluation of cyanobacterial cells removal and lysis by ultrafiltration. *Separation and Purification Technology*, 70 (2010) 345–353.

12. S. Sorlini, F. Gialdini, and C. Collivignarelli, Removal of cyanobacterial cells and microcystin-LR from drinking water using a hollow fiber microfiltration pilot plant. *Desalination*, 309 (2013) 106–112.

13. J.C. Crittenden, R.R. Trussell, D.W. Hand, K.J. Howe, and G. Tchobanoglous, *Water Treatment: Principles and Design*, 2nd ed. Hoboken, NJ: Wiley, 2005.

14. J. Lee and H.W. Walker, Effect of process variables and natural organic matter on removal of microcystin-LR by PAC-UF. *Environmental Science and Technology*, 40 (2006) 7336–7342.

15. C. Donati, M. Drikas, R. Hayes, and G. Newcombe, Microcystin-LR adsorption by powdered activated carbon. *Water Research*, 28 (1994) 1735–1742.

16. G. Newcombe, *Removal of Algal Toxins from Drinking Water Using Ozone and GAC.* American Water Works Association Research Foundation, Denver, CO, 2002.

17. M. Campinas and M.J. Rosa, Removal of microcystins by PAC/UF. *Separation and Purification Technology*, 71 (2010) 114–120.

18. Y. Zhang, J. Tian, J. Nan, S. Gao, H. Liang, M. Wang, and G. Li, Effect of PAC addition on immersed ultrafiltration for the treatment of algal-rich water. *Journal of Hazardous Materials*, 186 (2011) 1415–1424.

19. M.B. Dixon, Y. Richard, L. Ho, C.W.K. Chow, B.K. O'Neill, and G. Newcombe, A coagulation–powdered activated carbon–ultrafiltration—multiple barrier approach for removing toxins from two Australian cyanobacterial blooms. *Journal of Hazardous Materials*, 186 (2011) 1553–1559.

20. M.B. Dixon, Y. Richard, L. Ho, C.W.K. Chow, B.K. O'Neill, and G. Newcombe, Integrated membrane systems incorporating coagulation, activated carbon and ultrafiltration for the removal of toxic cyanobacterial metabolites from *Anabaena circinalis. Water Science and Technology*, 63 (2011) 1405–1411.

21. J.A. Westrick, D.C. Szlag, B.J. Southwell, and J. Sinclair, A review of cyanobacteria and cyanotoxins removal/inactivation in drinking water treatment. *Analytical and Bioanalytical Chemistry*, 397 (2010) 1705–1714.

22. B.C. Hitzfeld, S.J. Hoger, and D.R. Dietrich, Cyanobacterial toxins: removal during drinking water treatment, and human risk assessment. *Environmental Health Perspectives*, 108 (2000) 113–122.

23. V.K. Sharma, T.M. Triantis, M.G. Antoniou, X. He, M. Pelaez, C. Han, W. Song, K.E. O'Shea, A.A. de la Cruz, T. Kaloudis, A. Hiskia, and D.D. Dionysiou, Destruction of microcystins by conventional and advanced oxidation processes: a review. *Separation and Purification Technology*, 91 (2012) 3–17.

24. E. Rodríguez, G.D. Onstad, T.P.J. Kull, J.S. Metcalf, J.L. Acero, and U. von Gunten, Oxidative elimination of cyanotoxins: comparison of ozone, chlorine, chlorine dioxide and permanganate. *Water Research*, 41 (2007) 3381–3393.

25. G.D. Onstad, S. Strauch, J. Meriluoto, G.A. Codd, and U. von Gunten, Selective oxidation of key functional groups in cyanotoxins during drinking water ozonation. *Environmental Science and Technology*, 41 (2007) 4397–4404.

26. A.R. Shawwa and D.W. Smith, Kinetics of microcystin-LR oxidation by ozone. *Ozone: Science and Engineering*, 23 (2001) 161–170.

27. J. Rositano, B.C. Nicholson, and P. Pieronne, Destruction of cyanobacterial toxins by ozone. *Ozone: Science and Engineering*, 20 (1998) 223–238.

28. H.-F. Miao, F. Qinb, G.-J. Tao, W.-Y. Taoc, and W.-Q. Ruan, Detoxification and degradation of microcystin-LR and -RR by ozonation. *Chemosphere*, 79 (2010) 355–361.

29. S. Brooke, G. Newcombe, B. Nicholson, and G. Klass, Decrease in toxicity of microcystins LA and LR in drinking water by ozonation. *Toxicon*, 48 (2006) 1054–1059.

30. J. Rositano, G. Newcombe, B. Nicholson, and P. Sztajnbok, Ozonation of NOM and algal toxins in four treated waters. *Water Research*, 35 (2001) 23–32.

31. P.T. Orr, G.J. Jones, and G.R. Hamilton, Removal of saxitoxins from drinking water by granular activated carbon, ozone and hydrogen peroxide—implications for compliance with the Australian drinking water guidelines. *Water Research*, 38 (2004) 4455–4461.

32. X. Chen, B. Xiao, J. Liu, T. Fang, and X. Xu, Kinetics of the oxidation of MCRR by potassium permanganate. *Toxicon*, 45 (2005) 911–917.

33. E. Rodríguez, M.E. Majado, J. Meriluoto, and J.L. Acero, Oxidation of microcystins by permanganate: reaction kinetics and implications for water treatment. *Water Research*, 41 (2007) 102–110.

34. E. Rodríguez, A. Sordo, J.S. Metcalf, and J.L. Acero, Kinetics of the oxidation of cylindrospermopsin and anatoxin-a with chlorine, monochloramine and permanganate. *Water Research*, 41 (2007) 2048–2056.

35. K. Tsuji, S. Naito, F. Kondo, N. Ishikawa, M.F. Watanabe, M. Suzuki, and K.-I. Harada, Stability of microcystins from cyanobacteria: effect of light on decomposition and isomerization. *Environmental Science and Technology*, 28 (1994) 173–177.

36. R.K. Chiswell, G.R. Shaw, G. Eaglesham, M.J. Smith, R.L. Norris, A.A. Seawright, and M.R. Moore, Stability of cylindrospermopsin, the toxin from the cyanobacterium, *Cylindrospermopsis raciborskii*: effect of pH, temperature, and sunlight on decomposition. *Environmental Toxicology*, 14 (1999) 155–161.

37. K. Tsuji, T. Watanuki, F. Kondo, M.F. Watanabe, S. Suzuki, H. Nakazawa, M. Suzuki, H. Uchida, and K.-I. Harada, Stability of microcystins from cyanobacteria—II. Effect of UV light on decomposition and isomerization. *Toxicon*, 33 (1995) 1619–1631.

38. R.-P. Qiao, N. Li, X.-H. Qi, Q.-S. Wang, and Y.-Y. Zhuang, Degradation of microcystin-RR by UV radiation in the presence of hydrogen peroxide. *Toxicon*, 45 (2005) 745–752.

39. W. Song, T. Xu, W.J. Cooper, D.D. Dionysiou, A.A.d.l. Cruz, and K.E. O'Shea, Radiolysis studies on the destruction of microcystin-LR in aqueous solution by hydroxyl radicals. *Environmental Science and Technology*, 43 (2009) 1487–1492.

40. C. Xiaoliang, S. Honglan, C.D. Adams, T. Timmons, and M. Yinfa, Effects of oxidative and physical treatments on inactivation of *Cylindrospermopsis raciborskii* and removal of cylindrospermopsin. *Water Science and Technology*, 60 (2009) 689–697.

41. X. He, G. Zhang, A.A. de la Cruz, K.E. O'Shea, and D.D. Dionysiou, Degradation mechanism of cyanobacterial toxin cylindrospermopsin by hydroxyl radicals in homogeneous UV/H_2O_2 process. *Environmental Science and Technology*, 48 (2014) 4495–4504.

42. W. Song, S. Yan, W.J. Cooper, D.D. Dionysiou, and K.E. O'Shea, Hydroxyl radical oxidation of cylindrospermopsin (cyanobacterial toxin) and its role in the photochemical transformation. *Environmental Science and Technology*, 46 (2012) 12608–12615.

43. A. Afzal, T. Oppenländer, J.R. Bolton, and M.G. El-Din, Anatoxin-a degradation by advanced oxidation processes: vacuum-UV at 172 nm, photolysis using medium pressure UV and UV/H_2O_2. *Water Research*, 44 (2010) 278–286.

44. K. Nakata and A. Fujishima, TiO_2 photocatalysis: design and applications. *Journal of Photochemistry and Photobiology C: Photochemistry Reviews*, 13 (2012) 169–189.

45. O. Legrini, E. Oliveros, and A.M. Braun, Photochemical processes for water treatment. *Chemical Reviews*, 93 (1993) 671–698.

46. G.S. Shephard, S. Stockenström, D. De Villiers, W.J. Engelbrecht, E.W. Sydenham, and G.F.S. Wessels, Photocatalytic degradation of cyanobacterial microcystin toxins in water. *Toxicon*, 36 (1998) 1895–1901.

47. L.A. Lawton, P.K.J. Robertson, B.J.P.A. Cornish, I.L. Marr, and M. Jaspars, Processes influencing surface interaction and photocatalytic destruction of microcystins on titanium dioxide photocatalysts. *Journal of Catalysis*, 213 (2003) 109–113.

48. P.J. Senogles, J.A. Scott, G. Shaw, and H. Stratton, Photocatalytic degradation of the cyanotoxin cylindrospermopsin, using titanium dioxide and UV irradiation. *Water Research*, 35 (2001) 1245–1255.

49. Y. Tominaga, T. Kubo, and K. Hosoya, Surface modification of TiO_2 for selective photodegradation of toxic compounds. *Catalysis Communications*, 12 (2011) 785–789.

50. M.G. Antoniou, A.A. de la Cruz, and D.D. Dionysiou, Degradation of microcystin-LR using sulfate radicals generated through photolysis, thermolysis and e– transfer mechanisms. *Applied Catalysis B: Environmental*, 96 (2010) 290–298.

51. X. He, A.A. de la Cruz, and D.D. Dionysiou, Destruction of cyanobacterial toxin cylindrospermopsin by hydroxyl radicals and sulfate radicals using UV-254 nm activation of hydrogen peroxide, persulfate and peroxymonosulfate. *Journal of Photochemistry and Photobiology A: Chemistry*, 251 (2013) 160–166.

52. D. Chen, Applications of Ultrasound in Water and Wastewater Treatment, in: D. Chen, S.K. Sharma, and A Mudhoo (Eds.), *Handbook on Applications of Ultrasound: Sonochemistry for Sustainability*. Boca Raton, FL: CRC Press, 2011, Chap. 15.

53. D. Chen, S.K. Sharma, and A. Mudhoo (Eds.), *Handbook on Applications of Ultrasound: Sonochemistry for Sustainability*. Boca Raton, FL: CRC Press, 2011.

54. W. Song, A.A. de la Cruz, K. Rein, and K.E. O'Shea, Ultrasonically induced degradation of microcystin-LR and -RR: identification of products, effect of pH, formation and destruction of peroxides. *Environmental Science and Technology*, 40 (2006) 3941–3946.

55. W. Song, T. Teshiba, K. Rein, and K.E. O'Shea, Ultrasonically induced degradation and detoxification of microcystin-LR (cyanobacterial toxin). *Environmental Science and Technology*, 39 (2005) 6300–6305.

56. P. Gajdek, Z. Lechowski, T. Bochnia, and M. Kępczyński, Decomposition of microcystin-LR by Fenton oxidation. *Toxicon*, 39 (2001) 1575–1578.

57. E.R. Bandala, D. Martínez, E. Martínez, and D.D. Dionysiou, Degradation of microcystin-LR toxin by Fenton and photo-Fenton processes. *Toxicon*, 43 (2004) 829–832.

58. Y. Zhong, X. Jin, R. Qiao, X. Qi, and Y. Zhuang, Destruction of microcystin-RR by Fenton oxidation. *Journal of Hazardous Materials*, 167 (2009) 1114–1118.

59. F. Al Momani, Degradation of cyanobacteria anatoxin-a by advanced oxidation processes. *Separation and Purification Technology*, 57 (2007) 85–93.

60. B.-L. Yuan, J.-H. Qu, and M.-L. Fu, Removal of cyanobacterial microcystin-LR by ferrate oxidation–coagulation. *Toxicon*, 40 (2002) 1129–1134.

8

Summary and Conclusions

Mitigating the impacts of freshwater harmful algal blooms (HABs) and HAB toxins on humans and ecosystems is a challenging and complex problem. The issues are multifaceted and require an understanding of the causes and drivers of HABs, bloom dynamics, toxicology and impacts of HABs on human health, ecotoxicity, policy and regulatory instruments to reduce HABs, and the engineering design of conventional or advanced drinking water treatment processes. This book has provided an overview of the state of knowledge of these different facets.

It is clear our understanding of the causes, impacts, and approaches for the mitigation of HABs in freshwater systems has increased substantially in recent years. The fact that HABs are detected with greater frequency is at least partially because of our greater understanding of the problem. Regional approaches for reducing nutrient inputs to freshwater lakes and reservoirs are being developed and implemented, with varying degrees of success. Our better understanding of the fate of HABs and HAB toxins in water treatment plants has provided insight into how to better optimize conventional treatment processes to minimize or reduce toxin concentrations in finished water. New, advanced treatment processes, such as membranes and advanced oxidation processes (AOPs), potentially provide more effective treatment of HAB toxins, but problems remain to be solved in implementing these technologies for the treatment of HAB toxins in surface water.

Although much has been learned over the last decade regarding the causes, impacts, and mitigation of freshwater HABs, questions and needs remain. Understanding and reducing the impacts of freshwater HABs and HAB toxins, for example, require enhanced capabilities in the area of detection and prediction. The detection of HAB toxins is complicated by the fact that many different classes of toxins exist, and that within a given class of toxin (e.g., microcystins), many different congeners may be present. Identifying and quantifying the concentration of many different types of HAB toxins and congeners are time consuming and expensive. Also, water quality conditions during HABs present a complex matrix for detailed analysis because of the presence of large concentrations of natural organic matter. New sensing technologies are needed that are capable of quantifying HAB toxins quickly, in situ, and in real time. New, more effective sensing and detection approaches will facilitate larger monitoring efforts, both spatially and temporally, which will aid in establishing a better understanding of the factors controlling bloom formation. One challenge in developing new sensing and analytical techniques for the detection of HAB toxins, however, is the lack of analytical standards. Although progress has been made in the last 10 years in the availability of standard compounds, additional research is needed, as outlined in a number of scientific reports and publications [1–3].

The real-time, *in situ* detection of HAB toxins supports the development of HAB early warning systems. To develop better early warning systems, however, also requires more realistic models of bloom formation and toxin release. To develop such models and warning systems, better understanding of bloom dynamics is needed. Many HABs occur in large, relatively shallow, water bodies, which presents a challenge for hydrodynamic modeling because of the large computational requirements. Coupling the ecology to hydrodynamic models with the myriad factors and organisms important in bloom formation and maintenance is a challenge. The successful development of models will not only provide a way to provide early warning (on the order of days) for public health officials and lake managers but also help inform our understanding of the causes contributing to HABs and HAB toxin release. This better understanding will help in the formulation of effective policies and practices for reducing the prevalence of HABs in the future.

The full impacts of HABs and HAB toxins are still not fully understood. Better understanding of the ecology of HABs will help elucidate both the short- and long-term impacts of HABs on ecosystems and people. A major question, for example, concerns the effects of long-term (days to months to years) exposure to sublevel concentrations of HAB toxins on food webs as well as humans. Also, the factors and causes that control the production and release of HAB toxins during blooms are still rather poorly understood. Retrospective analyses of past HABs provide an opportunity to shed light on some of these questions, provided sufficient data collection and monitoring occurred during the event. Studies aimed at understanding the molecular biology of HABs and toxin production will also shed light on these questions, especially with respect to the factors that control HAB proliferation, gene expression, and toxin production and release.

A better understanding of the specific impact of HAB toxins on humans and animals is also needed to more clearly identify both the short- and long-term impacts of HABs. This knowledge will facilitate the development of improved human and ecological risk assessments of HABs and support the development of limits or guidelines for drinking water and natural aquatic systems. The lack of toxicological data is one factor hindering progress in this area, as is the need to better understand the fate and transport of HAB toxins in natural aquatic systems. Few studies, for example, have been conducted to elucidate the significance of physical, chemical, and biological processes in affecting the transport of specific HAB toxins in natural systems. A variety of processes may influence HAB toxin transport, including adsorption, photochemical degradation, and biodegradation, to name a few. The relative importance of these and other processes in controlling toxin fate is only poorly understood.

The reduction of HABs will largely depend on successful efforts to reduce the input of nutrients to impacted lakes and reservoirs. This will likely require new policies and practices to reduce both point and nonpoint sources of nutrients. As currently recognized, best management practices are needed to greatly reduce or eliminate nutrient inputs to impacted surface water bodies. Given the increasing occurrence of HABs, current efforts to develop and implement BMPs must be strengthened. New agricultural practices should also be examined to significantly reduce the need for fertilizer and reduce nonpoint inputs. The runoff of fertilizer to streams and rivers represents a cost to the agricultural industry. Presumably, the more efficient use of fertilizer will not only improve water quality but also minimize the cost of fertilizer application. Reducing the input of nonpoint sources of nutrients is preferable to other approaches as it reduces the need for downstream mitigation of HABs. In some cases, surface water nutrient levels are affected by the subsurface flux of nutrients (typically originating from ineffective on-site wastewater systems) to surface water. Reduction of this subsurface flux requires implementation of new, advanced on-site wastewater systems that are capable of reducing nutrient levels, as typical "septic tanks" provide little or no removal of nutrients. Programs or policies to reduce the runoff of residential lawn care chemicals are also needed in many cases. The creation of integrated hydrodynamic-ecological-social models and decision tools will facilitate the development of effective policies for reducing nutrient inputs and preventing HABs.

When efforts to prevent HABs fail, efforts are needed to mitigate the magnitude, extent, and impact of HABs. Given the many challenges in reducing nutrient levels in lakes and reservoirs, as evidenced by the lack of progress in this area, more research is needed to provide better approaches for the in situ control of HABs. As discussed in a previous chapter, a number of approaches, including destratification, flushing, altering nutrient ratios, and adsorption/coagulation, have been explored for altering physical, chemical, or biological conditions to mitigate HABs in a particular water body. More work in this area is clearly needed to provide lake managers more tools in combating HABs, either during or shortly before a bloom is expected to occur. A better understanding of bloom dynamics will support this goal, as the successful implementation of physical, chemical, or biological alteration of a reservoir or lake for HAB mitigation necessitates a thorough understanding of lake ecology and factors promoting HAB formation.

When prevention, control, or mitigation efforts fail, drinking water treatment plants must be prepared to deal with HABs and the toxins they produce. Water treatment plants provide perhaps the last line of defense to protect public health from the deleterious impacts of HABs and HAB toxins. As discussed previously, conventional treatment processes such as coagulation, sedimentation, filtration, and chlorination must be optimized to reduce HAB cells and prevent the release of HAB toxins during treatment. As a solid separation process, coagulation and sedimentation can effectively reduce the concentration of HAB cells. However, care must be taken during treatment not to disrupt cells during solids removal and release dissolved toxins. Chlorination, which is also a typical conventional treatment approach, can be effective for the removal of some HAB toxins. Effective chlorination, however, typically requires long contact times or high dosages not typical of most treatment plants. Chlorination can also disrupt intact HAB cells, leading to toxin release.

Although conventional treatment processes can be effective at removing HAB cells and achieving partial toxin removal, better understanding of how specific unit operations may disrupt cells and release toxins during treatment is needed. More full-scale studies are needed to elucidate the fate of HABs and HAB toxins in conventional water treatment plants. These studies will enable water treatment plant operators to better optimize existing treatment practices to maximize cell and toxin removal and minimize cell disruption and toxin release. Additional insight is also needed in understanding how different congeners are removed during conventional treatment processes, as most research to date has focused on only a limited number of HAB toxins. During bloom events, source water conditions may change significantly compared to other operating periods, which may also lead to operational challenges in optimizing HAB cell and toxin removal. Research is needed to identify how bloom-related changes in water quality may have an impact on cell and toxin removal.

New technologies offer promise in achieving better removal of HAB cells and associated toxins. New technologies generally need to meet a number of criteria to be feasible alternatives to existing processes, including (1) be effective for a broad range of HAB toxins, as well as other similar classes of contaminants (e.g., trace organics); (2) be economical; (3) do not disrupt existing treatment operations or make existing treatment operations more expensive; and (4) do not generate secondary metabolites of either greater toxicity or unknown toxicity. Technologies such as membrane processes and AOPs have the potential to meet these criteria, but additional research is needed. Membrane processes, for example, have been effective for broad classes of contaminants (including HAB toxins), both organic and inorganic, but are prone to membrane fouling. Membrane fouling is particularly an issue in surface treatment during HABs given the high organic loading during such events. The costs of membrane processes have decreased in recent years but are still relatively high compared to more traditional processes. New chemical approaches, such as AOPs, also offer promise, but more work is needed to reduce the cost of these processes for full-scale treatment and to better understand the formation of secondary by-products.

Although the costs to improve drinking water treatment or implement best management practices for nutrient reduction are often cited as hurdles to solving the HAB problem, comparatively less research has explored the socioeconomic costs of not addressing the problem. Better understanding of the full costs to society of HABs is needed. Research exploring the impact of HABs on tourism, recreation, fisheries, and industries, for example, is needed. The public health costs of HABs are also not well known, especially with respect to the health effects of long-term, low-level exposure to HAB toxins. HABs also affect the ability of streams, lakes, and reservoirs to provide critical ecosystem services, including "supporting," "regulating," "provisioning," and "cultural" services. By disrupting normal ecosystem processes, for example, HABs have an impact on the ability of lakes and reservoirs to provide fresh water and reduce the aesthetic value of water resources, and they have an impact on nutrient cycling and primary productivity.

In the end, policy makers, lake managers, and engineers must make decisions to reduce or mitigate the impacts of freshwater HABs and HAB toxins on humans and ecosystems. This decision making will be supported by better understanding of the causes of HABs, the toxicology and impacts of HABs on human health and environment, the effectiveness of policy or regulatory approaches to reduce HABs, and the engineering design of conventional or advanced drinking water treatment processes for reducing HAB toxins.

References

1. M. Burch, Effective doses, guidelines and regulations, in: K.H. Hudnell (Ed.), *Cyanobacterial Harmful Algal Blooms: State of the Science and Research Needs*. New York: Springer, 2008, pp. 833.
2. H.K. Hudnell, The state of US freshwater harmful algal blooms assessments, policy and legislation. *Toxicon*, 55 (2010) 1024–1034.
3. C.B. Lopez, E.B. Jewett, Q. Dortch, B.T. Walton, and H.K. Hudnell, *Scientific Assessment of Freshwater Harmful Algal Blooms*. Interagency Working Group on Harmful Algal Blooms and Human Health of the Joint Subcommittee on Ocean Science and Technology, Washington, DC, 2008.

Index